Elementary
Statistics

THIRD EDITION

JANET T. SPENCE
University of Texas at Austin

JOHN W. COTTON
University of California, Santa Barbara

BENTON J. UNDERWOOD
Northwestern University

CARL P. DUNCAN
Northwestern University

PRENTICE-HALL, INC., *Englewood Cliffs, New Jersey*

Library of Congress Cataloging in Publication Data

Main entry under title:

Elementary statistics.

First ed. by B. J. Underwood and others.
Includes index.
1. Statistics. I. Spence, Janet Taylor.
II. Underwood, Benton J., date Elementary statistics.
HA29.E43 1976   519.5   75–41610
ISBN 0–13–260109–5

THE CENTURY PSYCHOLOGY SERIES
Richard M. Elliott / Gardner Lindzey / Kenneth MacCorquodale
Editors

Printed in the United States of America

10   9   8   7   6

Prentice-Hall International, Inc., *London*
Prentice-Hall of Australia, Pty. Ltd., *Sydney*
Prentice-Hall of Canada, Ltd., *Toronto*
Prentice-Hall of India Private Limited, *New Delhi*
Prentice-Hall of Japan, Inc., *Tokyo*
Prentice-Hall of Southeast Asia (PTE.) Ltd., *Singapore*

Contents

Preface

Undergraduate students in psychology and education come to their first course in statistics with diverse expectations and backgrounds in mathematics. Some have considerable formal training and quantitative aptitude and look forward to learning about statistics. Others—perhaps the majority, even of those who aspire to postgraduate study—are less certain in their quantitative skills and regard a course in statistics as a necessary evil if they are to understand or to perform research in their chosen field but an evil nonetheless.

This third edition of *Elementary Statistics*, like its predecessors, is directed primarily at the latter audience. It was written with the conviction that statistical concepts can be described simply without loss of accuracy and that understanding of statistical techniques as research tools can be effectively promoted by discussing them within the context of their application to concrete data rather than as pure abstractions. Further, its contents are limited to those statistical techniques that are widely used in the psychological literature and to the principles underlying them.

The changes that have been made in this edition reflect both the results of our teaching experience and the increasing prominence being given by statisticians to certain topics. Thus our discussions of some procedures, particularly those in the realm of descriptive statistics, which students grasp easily, have been shortened or rearranged. The chapter on percentiles, for example, has been omitted, the students being introduced to the topic in the discussion of cumulative frequency polygons and to methods of computation in the section of the chapter on measures of central tendency in which the median is discussed. The treatment of other topics has been expanded. Greater emphasis has been placed on sampling theory, hypothesis testing, and the notion of statistical power.

A new chapter, on the elementary principles of probability, has been included. In addition to teaching the students about concepts of value in their right, the material in this chapter serves as an introduction to the discussions of hypothesis testing that follow in later chapters and, even more particularly, to the discussion of the binomial distribution that has been added to the chapter in which the normal probability distribution is presented.

Finally, several instructional aids have been added. A list of terms and symbols to review concludes each chapter. Major statistical formulas in each chapter have been numbered for easy reference, and the complete set of formulas, identified by number and the page in the text on which they appear, are reproduced in an appendix.

Responsibility for preparation of the present edition was assumed by the first two authors. Our goal was to remain faithful to the style and intent of the earlier editions.

Supplementing the text is a separate *Workbook* for *Elementary Statistics*, prepared by the text's senior author. The *Workbook* contains exercises for each chapter in the text, designed to give the student practice in applying statistical techniques to research data and to amplify understanding of statistical principles.

We are grateful to a number of authors and publishers for granting us permission to abridge or reproduce their materials: The American Psychological Association for its several journals; the late Carl Murchison for the *Journal of Psychology*; the Williams & Wilkins Company for the *Journal of Comparative Psychology*. Table A in Appendix II is included by courtesy of the Cambridge University Press and the Houghton Mifflin Company, Table B, by courtesy of the Houghton Mifflin Company and Table C by permission of D. B. Owen and Addison-Wesley. We are grateful to the literary executor of the late Sir Ronald A. Fisher, F.R.S., and to Dr. Frank Yates, F.R.S., and to Longman Group Ltd., London, for permission to abridge Tables III, IV, and VI from their book, *Statistical Tables for Biological, Agricultural, and Medical Research* (6th ed., 1974). Data from these tables make up Table G and a portion of D. The remainder of Table D is reprinted with the permission of the Iowa State University Press and Professor George Snedecor. Table E is reprinted by courtesy of the Institute of Mathematical Statistics and the late Professor E. G. Olds and Table F by courtesy of the editor of *Biometrika*. Table H is adapted from Professor Frank Wilcoxon's book, *Some Rapid Approximate Statistical Procedures,* with the permission of the publisher, the American Cyanamid Company. Table I is adapted from *Biometrika Tables for Statisticians,* Vol. 1 (3rd ed.) edited by Professors E. S. Pearson and H. O. Hartley, with the permission of the Biometrika trustees. Finally, we are grateful to the editor of *Biometrika* and to Holt, Rinehart and Winston, Inc., for allowing us to adapt certain procedures presented in Chapter 12, and to John Wiley and Sons, Inc., for allowing us to reproduce certain materials appearing in Chapter 14.

We thank Richard Houang for his painstaking efforts in checking formulas, computational examples, and transcriptions of table entries and for his suggestions about the text. Thanks are also due to Andrea Weiland for her skillful proofreading of text and tables.

JTS

JWC

Numerical Quantification and Types of Data

1

Suppose you are driving downtown to do some shopping. You remember your tires are low, so you stop at a service station and have them brought up to 24 pounds, and while you're at it you put 10 gallons in your gas tank. In the course of your shopping you have a lumberyard cut three 5-foot planks for some shelves you are building, order a sweater in your size, buy a 40-watt light bulb and select a half pound of mushrooms. On your way home you glance at your watch to see if you'll make your favorite TV program.

All of these commonplace activities depend on *quantification*—the assignment of *numbers* to objects or events to describe their properties. If you had attempted to express your requirements without numbers, you would have found yourself using vague verbal descriptions, gesturing a lot, and failing to communicate your exact needs. Numbers allow you to describe phenomena with considerably more precision than purely verbal expressions. If numerical quantification had never been devised, it is doubtful that cars, light bulbs, watches, and the like would even have existed.

In scientific research we attempt to describe the properties of objects and events in the universe around us and the regularities of occurrence among them. Quantification is at the heart of the scientific enterprise, and the basic data resulting from scientific research are almost invariably a set of numbers. Having collected the data, we apply statistical methods to describe and interpret them.

Let us look at a hypothetical psychological investigation to see some of the statistical challenges. Many students consistently become so anxious and upset during exams that they don't do well. "Systematic desensitization" has been suggested as a therapeutic procedure for reducing examination jitters. Students first are trained in muscular relaxation and then, while in a relaxed, nonanxious state, practice imagining increasingly anxiety-provoking scenes related to studying and exam-taking.[1] We decide to test the technique and put ten highly anxious volunteers from an introductory psychology class through a systematic desensitization program. An equal number of terrified students who have volunteered for therapy are not treated and serve as a control group. At semester's end we get the final exam scores of all twenty students from the psychology instructor, as shown in Table 1.1.

[1] See Johnson, S., and Sechrest, L. Comparison of desensitization and progressive relaxation in treating test anxiety. *Journal of Consulting and Clinical Psychology*, 1968, 32, 280–286, as an example of an investigation of systematic desensitization in the treatment of text anxiety; and Wolpe, J. *Psychotherapy by reciprocal inhibition*, Stanford, Calif., Stanford University Press, 1958, for a description of the theory of systematic desensitization.

Table 1.1 Results of a hypothetical experiment showing final examination scores of treated and untreated students who are anxious about tests.

Therapy Group	No-Therapy Group
74	79
81	68
77	75
89	59
72	71
69	91
77	63
84	82
64	76
79	67

Did the treatment have beneficial effects? We cannot even begin to answer the question until we have brought some sort of order into these jumbled sets of scores and described them in some way. For example, what was the typical or average score in each group? In either group, how variable were the scores—that is, were they very similar or did they range all over the place?

Suppose we found that the average treated student did do better than the average untreated student. Does this mean that the therapy worked—or that the treated students just happened by chance to do better? In other words, if the experiment were repeated over and over with different groups of students, is it likely that the results would turn out the same way?

Simply staring at the subjects' scores will not bring us satisfactory answers to these questions. Obviously, we need methods of extracting information from our raw data. We need some way to discover and describe the relationships hidden in masses of numbers. This is where statistics comes in. By performing arithmetic and algebraic manipulations with our scores we can abstract group trends and relationships and express them in a precise, accurate manner. With further statistical operations, we can make inferences or guesses about the characteristics of still larger groups—for example, about the reactions of introductory psychology students in general to our therapeutic procedures. Only with such information can we throw light on the type of problem our illustrative experiment was originally designed to investigate.

In subsequent chapters we first will consider a set of *descriptive statistics*, then take up *inferential statistics*—techniques for making guesses about the properties of large groups of individuals based on the data from the samples of individuals we have been able to observe. Before proceeding, however, we must discuss the major types of numerical information we will be encountering: *categorical*, *measurement*, and *rank-order* data.

CATEGORICAL DATA

In your statistics class, how many men and women are enrolled? How many of you are freshmen, sophomores, juniors, or seniors? How many of you have been to Europe, smoked pot, own a car, have shoes on, are blondes, brunettes, or redheads?

Each of these questions involves a set of two or more mutually exclusive *categories*: male *or* female; car owner *or* not; blonde, brunette, *or* redhead; and so on. An individual either does or does not have the property specified by each category within a given set (that is, his "score" is either 0 or 1), and since the categories are mutually exclusive, each individual can be assigned to one and only one of them. More than one set of categories can, of course, be used simultaneously in classifying an individual case; for example, we can identify Samantha as a redheaded freshman girl who came to class barefoot.

It is easy to order categorical data so that we can describe the results of our observations. We simply sort the cases, placing each in its proper category, then *count* the number in each category and present the sums. For convenience we may translate the sums into percents. Table 1.2, for example, gives the percent of groups of men and of women in three different years who said "Yes," "No," or "Can't say" about whether they were afraid to walk alone at night. The table lets us see that in each year, more women than men reported being afraid, but that both sexes were becoming more fearful over the three-year period.

NUMERICAL DATA

Measurement

Instead of inquiring about whether you're afraid to go out alone at night or about the sex or hair color of members of your class, we could ask each of you about your height, how fast you can hop around the room on one foot, or what score you think you'll make on your first statistics exam. The property specified in each instance exists along a dimension in *various degrees*. We describe each case by assigning a *number* that specifies *how much* of the property the individual has. These numbers we refer to as *measurement* or *metrical* data.

Table 1.2   Percent of men and of women afraid to walk alone
at night in 1967, 1968, and 1972.

	Women			Men		
Year	Yes	No	Can't Say	Yes	No	Can't Say
1967	44	53	3	16	82	2
1968	50	47	3	19	79	2
1972	61	39	0	22	77	1

SOURCE: Adapted from Executive Office of the President: Office of Management
and Budget. *Social Indicators, 1973*. U.S. Government Printing Office, Washington,
D.C., 1973.

Discrete vs. Continuous Variables.   Measurable properties are usually
continuous. That is, objects or events may fall at *any* point along an unbroken
scale of values, running from high to low. In measuring the weight of objects,
for example, we may have such values as 4.2681 grams, or 68.723584 grams,
or any positive value we can think of. Similarly, time is continuous, as is the
distance between geographic locations.

Discrete variables, in contrast, fall only at certain points along a scale.
If we tossed 50 pennies, for example, we could determine how many heads
came up, but the resulting number could only be an integer: 0, 1, 2, 3, and
so on; fractional numbers such as 23.278 heads could not occur. As another
example of a discrete measure, students' scores on a 20-item true-false test
would be whole numbers; they would fall only at certain points along a
scale running from 0 to 20.

In practice, the distinction between the two types of measures breaks
down. When the property is continuous, we can never measure objects
exactly but only within the limits of a specified degree of accuracy. Depending on our purpose, for example, we might be satisfied to specify an
object's age to the nearest century, the nearest year, the nearest hour, or the
nearest millionth of a second. Whatever the degree of accuracy we attempt,
the resulting numbers are still discrete. If we measured time to the nearest
tenth of a second, for example, numbers such as .1, .3, or .8 second would be
reported but not .12, .32, or .86. If we measured to the nearest hundredth
of a second, we might have .08 and .36, but not .084 or .362. Because most
measurable properties are continuous, each number we assign to each object
or event typically represents a range or *interval* of values, rather than an
exact scale point. If length is reported to the nearest centimeter, for example,
objects described as being 14 centimeters long are understood to fall someplace in a range extending from 13.5 to 14.5 and not all to be exactly
14.000...

A Note about Rounding. It is not uncommon to record observations with greater numerical precision than we wish to report, or to encounter other situations in which we want to round the numbers we obtain. We might have recorded our observations using two decimal places, for example, but want to report data only to the first decimal place or in whole numbers. The traditional rule is to round to the *nearest* decimal or whole number. The value 4.38, for example, would be rounded *down* to 4 if a whole number were required or rounded *up* to 4.4 if the first decimal place were to be retained. What if the number falls at the border of two adjacent intervals? For example, is 9.5 rounded to 9 or to 10? Is 2.65 rounded to 2.6 or 2.7? A convention devised for handling such situations is to round to the nearest *even* number, so that 9.5 is rounded up to 10 and 2.65 is rounded down to 2.6. This convention is intended to avoid systematic bias when a whole group of numbers is to be rounded, since it can be anticipated that about half the numbers falling on the borderline will be assigned to the lower interval and half to the upper interval.

Rank-Order Data

On occasion numbers that directly indicate how much of a given property each individual possesses are impossible or inconvenient to determine. We may instead assign *ranks* that indicate the individuals' position relative to each other. In a foot race, for example, the contestants might not be timed but instead ranked from fastest to slowest according to the order in which they crossed the finish line. Or sportswriters may be asked to rank their choices for the top twenty football teams in the country. Rank orders allow us to make more-than, less-than statements about pairs of objects (for example, Ohio State has a better team than Indiana) but not to indicate how great the difference is.

Having observed a group of objects or events in some manner, what are we to do with the resulting numbers? In the next chapter we consider measurement data and how to order them so that we may begin to describe them.

TERMS TO REVIEW

Quantification	Metrical data
Descriptive statistics	Continuous variables
Inferential statistics	Discrete variables
Categorical data	Interval of values
Measurement data	Rank-order data

Frequency
Distributions

2

In Chapter 1 we described measurement data: the assignment of numbers that specify *how much* of some property individuals have. We might measure, for example, the reaction times of a group of adults before and after the ingestion of a drug, the performance of young children on a task designed to determine their understanding of abstract moral principles, or the ego-strength of hospitalized schizophrenics as revealed in their responses in a standardized psychiatric interview. When a group of individuals is measured, rarely do all the measures turn out to be the same; the almost inevitable outcome is a set of numbers varying in value. In this chapter we will consider methods of ordering measurement data so that we may begin to comprehend and describe the performance of a group of differing individuals.

SIMPLE FREQUENCY DISTRIBUTIONS

One obvious method of bringing order into a collection of measures is to arrange them in order of magnitude from highest to lowest. When only a few measures are involved, this ordering is probably sufficient, but with a group of substantial size it is no longer immediately evident how many times a particular value occurs, or whether it occurs at all. An additional step in our procedure will give us this extra information. All possible values between the highest and lowest reported measures can be listed in one column and the number of individuals receiving each value in an adjacent column, as illustrated in Table 2.1.

The numbers in the table represent scores received by a group of 100 female college students on a scale assessing attitudes toward women's sex roles. (The higher the score, the greater the belief that the expectations society has for women and the rights and privileges granted to them ought to be the same as for men.) As can be seen, the first column is labeled "score." The order of highest to lowest seen in the "score" column has, by custom, become the standard arrangement and should always be followed. The second column, labeled "f" (for "frequency"), contains the number of cases, in this instance the number of students receiving each score. For example, three students received a score of 62 and five a score of 45. At the bottom of the last frequency (f) column you will observe the letter N, which is simply a symbol for the total number of cases in a group. In the present illustration N is 100. Such a table, in which all score values are listed in one column and the number of individuals receiving each score in the second, is called a *simple frequency distribution*.

Table 2.1   Simple frequency distribution of attitude scores for 100 female students.

Score	f	Score	f	Score	f	Score	f
70	1	59	3	48	1	37	1
69	1	58	4	47	1	36	3
68	3	57	3	46	2	35	2
67	1	56	4	45	5	34	0
66	3	55	6	44	2	33	3
65	1	54	2	43	1	32	3
64	2	53	5	42	3	31	1
63	1	52	3	41	4	30	2
62	3	51	2	40	2	29	0
61	2	50	4	39	4	28	1
60	2	49	3	38	0	N =	100

SOURCE: The data are a representative sample taken from the groups reported in Spence, J. T., Helmreich, R., and Stapp, J. A short version of the Attitudes toward Women Scale (AWS). *Bulletin of the Psychonomic Society*, 1973, 2, 219–220. Raw data were obtained by courtesy of the authors.

Arranging the measures into a simple frequency distribution makes description of group performance easier. In our example we find that certain scores stand out as occurring frequently and, more importantly, that the scores are heavily concentrated toward the high end of the scale—not too surprising in view of the vigor of the feminist movement among college women. Collecting scores into a frequency distribution allows us to see these characteristics of group performance more easily than would just listing scores in order.

GROUPED FREQUENCY DISTRIBUTIONS

In many instances, particularly those in which the *range* of the measures (number of score units from highest to lowest) is small, a frequency distribution of the type just described yields a compact table that shows the pattern formed by a set of scores. But when the measures are scattered over a wide range, the frequency distribution is long and bulky, as in the four long pairs of columns in Table 2.1. More important, so few cases fall at each score value that the group pattern is not too clear. We can, however, simultaneously make our distribution more compact and obtain a better idea of the overall pattern of scores by bringing together the single measures into a number of groups, each containing an *equal* number of score units. Table

2.2 does this with the attitude scores from Table 2.1, grouping them into blocks of three score units each. Each block or *class interval* is identified by its "class limits"—the highest and the lowest score in the interval. Thus, the top interval in the first column of Table 2.2 includes the scores 68, 69, and 70 and is identified as class 68–70. The next class consists of the three scores 65, 66, and 67, and so forth.

Table 2.2 Grouped frequency distribution of attitude scores for 100 female students, where $i = 3$.

Class Interval	f	Class Interval	f
68–70	5	44–46	9
65–67	5	41–43	8
62–64	6	38–40	6
59–61	7	35–37	6
56–58	11	32–34	6
53–55	13	29–31	3
50–52	9	26–28	1
47–49	5	N =	100

We can describe these intervals by saying that each has a *width* of three, since three score units are included in each. The symbol we will be using to indicate the width of the interval in any grouped frequency distribution is *i*. The second column again contains frequencies (*f*), but these are now the number of individuals falling into each class interval rather than the number receiving each separate score. The resulting distribution is an improvement over the previous one in showing group trends. We can now see quite clearly that the scores for the students tend to cluster near the high end of the scale and trail off in numbers toward the low end.

The degree of condensation obtained by grouping data is determined by the number of score units included in each class interval, since increasing the size of *i* decreases the number of class intervals. At the left of Table 2.3 the data are again represented, but *i* has been increased from 3 to 5 units, thus decreasing the number of intervals and making the table shorter. At the right of the table, $i = 10$ units, making the resulting grouped frequency distribution even more compact. A comparison of these frequency distributions, all derived from the same raw data, reveals an interesting fact. As the width of the class interval, *i*, increases, a pattern first emerges and later disappears as more and more scores are lumped together in fewer classes. We see that as far as obtaining information about group trends is concerned, a frequency distribution can be condensed too far.

A further point to be noted about grouping data is that the identity of the individual score is lost. In Table 2.3, for example, where $i = 10$, the topmost interval contains 18 women, but we do not know exactly what score each received; we know only that each person falls somewhere within the interval. As i becomes larger, the numerical value of the individual score becomes more and more obscured.

Table 2.3 Grouped frequency distribution of attitude scores, where $i = 5$ and 10.

Class Interval	f	Class Interval	f
66–70	9	61–70	18
61–65	9	51–60	34
56–60	16	41–50	26
51–55	18	31–40	19
46–50	11	21–30	3
41–45	15		N = 100
36–40	10		
31–35	9		
26–30	3		
	N = 100		

How Many Intervals?

Before a set of data can be converted into a grouped frequency distribution, we must decide how many class intervals to employ. This decision is largely arbitrary, depending on such factors as the number of cases, the range of measures, and the purpose for which the distribution is intended. As a general principle we might say that the classes should be few enough so that the resulting distribution reveals a group pattern, but not so few as to obscure a group pattern and cause great loss in the precision with which we can identify the individual score. A rule of thumb often given is that between 10 and 20 intervals should be employed, since most data can be adequately handled within these limits. However, in choosing the number of class intervals, or indeed in deciding whether the range of measures is so small that scores should not be grouped at all, we must examine each set of data individually.

How Many Units for the Class Interval?

Once the *approximate* number of class intervals has been chosen, we must determine the width of the class interval necessary to produce the desired number of intervals. We can roughly estimate the needed i by dividing the range of the scores (the highest minus the lowest score) by the number of intervals wanted. Thus, if a set of scores ran from 90 to 143 and about 15 intervals were to be employed, the approximate width would be 3.5 score units, since

$$\frac{143 - 90}{15} = 3.5$$

Since whole numbers are more convenient than fractions for widths of intervals, especially when the original measures are themselves integers, in this example we have a choice of either 3 or 4 units for i. In most situations there will be a similar choice, since dividing the range by the number of intervals wanted usually results in a fractional value.

On the basis of practical considerations, certain widths are preferable to others. When the choice is among widths smaller than 10, odd-sized widths (3, 5, 7, 9) are often more convenient than even widths. In the example above, therefore, we would select 3 as our i instead of 4. For interval sizes of 10 and above, multiples of 5 (10, 15, 20, and so forth) are usually selected.

Nothing is sacred about any of these widths; they just turn out to have certain practical advantages. As for the larger widths, we are a "ten-minded" people and are happier when dealing with multiples of 10 or 5 than with other numbers. For smaller widths there is a different reason for using odd numbers. We frequently will be using and referring to the "midpoint"— the middle score value in any interval. For example, 11 is the midpoint of the interval 10–12, 105 is the midpoint of the interval 103–107, 47.5 is the midpoint of 46–49. Note that in the first two examples i is an odd number (3 and 5 score units, respectively) and the midpoints are whole numbers (11 and 105); in the third example i is an even number (4 units) and the midpoint, 47.5, involves a fraction. These results are not coincidental: when i is an odd number, the midpoint is an integer, but when it is an even number, the midpoint of each interval is a fractional value. Since whole numbers are typically more convenient to deal with than fractions, odd-numbered widths are employed so that the midpoints will be whole numbers.

In the frequency distribution used as an illustration up to this point, the raw measures have been whole numbers. When such data are grouped, we note that there are gaps between the upper and lower limits of adjacent classes. Thus, if we had classes 10–12 and 13–15, the first class would go up to include (have an upper limit of) 12.0 and the next would start at (have a lower limit of) 13.0. Such values as 12.1 or 12.9, if they occurred, could not be assigned to either class, since they fall in the gap between 12.0 and 13.0. These class limits with gaps in between are identified as "apparent limits." By contrast, "real limits" take into account that the variable being measured is almost always continuous and that each discrete score value actually represents an *interval* of values. Real limits therefore extend the upper and lower apparent limits of adjacent classes equally until they meet at the point midway between them and close the gap. For example, the real limits of the class 10–12 cited above are 9.5–12.5, and those of the adjacent class 13–15 are 12.5–15.5. The point of this distinction between the two kinds of class limits will become apparent in later chapters, where we shall see that real rather than apparent limits are used in certain statistical operations.

TERMS AND SYMBOLS TO REVIEW

Score	Width of interval (i)
Frequency (f)	Range
Simple frequency distribution	Number of cases (N)
Grouped frequency distribution	Apparent limits
Class interval	Real limits
Class limits	

Graphic
Representations

3

Most of us have seen in newspapers and magazines graphs that present various types of statistical information: infant mortality rate in different countries, popularity of political candidates, the distribution of incomes, and so forth. Such graphs are used by popular writers in preference to tables of numbers because they are both more interesting and more easily interpreted than columns of figures. The graphic method is also employed extensively in technical articles, since the outstanding features of group performance are most readily grasped in this form.

GRAPHIC REPRESENTATIONS OF FREQUENCY DISTRIBUTIONS

We will discuss here two graphic methods of presenting frequency distributions: the frequency polygon and the cumulative frequency curve. Certain common-sense rules and ways of graphing data apply to both methods. These will be outlined after we look briefly at each type.

Frequency Polygon

Frequency distributions are most often presented graphically by means of a frequency polygon. Two frequency polygons, along with the distributions from which they were derived, are shown in Figure 3.1. We have plotted both distributions in a single graph, so that we can compare them more easily.

The frequency distributions in Figure 3.1 give the scores made by a group of 513 professional engineers and a group of 306 college freshmen on a test of engineering interest.[1] Since the number in each group is not the same, the frequency in each class interval is expressed in percentage to help us compare the different-sized groups. These distributions contain all the necessary information about the scores for the two groups, but we can see how much easier and faster it is to obtain this information by inspecting the polygons. We see first that the scores of the engineers are relatively concentrated around a score of 50. The scores of the freshmen, in contrast, are more variable and tend to be lower.

Let us now examine the technical features of a frequency polygon in

---

[1] Strong, E. K., Jr. Nineteen-year followup of engineer interests. *Journal of Applied Psychology*, 1952, 36, 65–74.

Engineers		Freshmen	
Midpoint	Percent	Midpoint	Percent
70	1	60	2
65	3	55	4
60	13	50	6
55	17	45	6
50	18	40	11
45	16	35	11
40	16	30	13
35	8	25	9
30	5	20	15
25	2	15	9
20	1	10	8
		5	4
		0	2

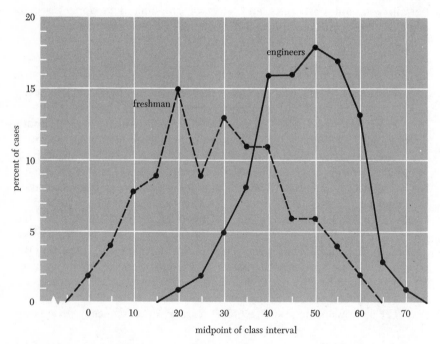

Figure 3.1 Frequency polygons showing engineering interest scores for adult engineers and college freshmen and the grouped frequency distributions from which they were drawn.

detail. There are two axes: the horizontal baseline, also known as the X axis or the abscissa, and the vertical or Y axis, often referred to as the ordinate. The scores (in this example grouped into class intervals) are represented along the X axis, while the Y axis represents frequency or percent of cases. Each axis is appropriately labeled, the numbers along the abscissa indicating the midpoints of the class intervals, those along the ordinate the percent of cases in each interval.

As can be seen, points are plotted over the midpoints of each interval, the height of each point being determined by the appropriate frequency or percent. The polygon is completed by connecting adjacent points with a straight line and dropping the lines to the baseline at either end to show the upper and lower intervals at which zero frequencies begin.

Cumulative Frequency Curve

The cumulative frequency curve, though used less often than the frequency polygon, is preferred when we are interested primarily in specifying the position of an individual or individual score in a distribution rather than the general form of the group performance. The results from many tests of abilities, personality attributes, and the like are reported with cumulative frequencies (or cumulative percents) and graphed in this fashion, since such scores are often used for individual diagnosis and evaluation. Before discussing this use of the cumulative curve, we will examine an example of a cumulative frequency curve and describe its construction.

In Chapter 2 a distribution was presented showing the scores obtained by female students on a questionnaire about women's roles. The distribution

Class Interval	f	cum f	cum %
66–70	3	80	100.0
61–65	4	77	96.2
56–60	6	73	91.2
51–55	9	67	83.8
46–50	10	58	72.5
41–45	13	48	60.0
36–40	14	35	43.8
31–35	10	21	26.2
26–30	8	11	13.8
21–25	3	3	3.8
N = 80			

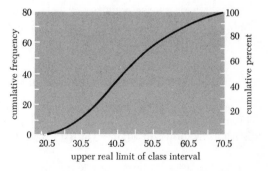

Figure 3.2 Cumulative curve of attitude scores for male students and the frequency distribution from which it was derived.

of scores[2] obtained from 80 *male* students on the same questionnaire is shown in Figure 3.2 along with the cumulative frequency curve derived from it. This curve, incidentally, has been "smoothed": that is, a continuous line has been drawn so that it comes as close as possible to all points. Scores (here class intervals) are indicated along the baseline of the graph and frequency along the Y axis, just as in a polygon. But there is a difference between the frequencies plotted in the polygon and those of the cumulative frequency curve. We are now dealing with cumulative frequencies: *the total number of cases earning a certain score or less.* As a demonstration, find the score 40.5 (an upper real class limit) along the baseline of Figure 3.2. Going up to the curve and across to the vertical axis, you discover that 35 people have scores of 40.5 or less. (Conversely, 45 have scores of 40.5 or *more.*) On the other hand, you will note that 73 individuals received a score of 60.5 or less (the upper real limit of class 56–60), indicating that an attitude score higher than this was relatively rare in the group of 80 male students.

To plot a cumulative frequency curve we first must derive the cumulative frequencies themselves from the distribution. In Figure 3.2 we find 3 individuals falling in the bottom interval and 8 in the second from the bottom; thus, 3 + 8 or 11 cases have scores of 30.5 or less. This sum, 11, is entered in the cumulative frequency column. For the next entry we find that 21 individuals have scores of 35.5 or below (3 + 8 + 10) or (11 + 10). By successive addition, from bottom to top, of the entries in the frequency column we get the cumulative frequencies. The entry in the topmost interval will be equal to the total number of cases, since all individuals have a score equal to or less than the upper limit of this class.

A cumulative frequency curve represents the graphing of a cumulative distribution and is carried out in a manner similar to the frequency polygon. In this method, however, the points are plotted directly above the *upper real limits* of each interval rather than the midpoint. The upper real limits are used because cumulative frequency represents the number or percent of cases that have scores equal to or below the value at the particular upper limit. The height of each point above the abscissa is, of course, determined by the appropriate cumulative frequency.

Cumulative Percentages and Percentiles

Cumulative frequency can also be translated into cumulative percent. This translation has been made in the right-hand column of the frequency distribution shown in Figure 3.2. The vertical axis at the right has also been

---

[2] The data were taken from a representative sample of the groups reported in Spence, J. T., Helmreich, R., and Stapp, J. A short version of the Attitudes toward Women Scale (AWS). *Bulletin of the Psychonomic Society*, 1973, 2, 219–220. Raw data were obtained by courtesy of the authors.

marked off in percents, so that we may read the curve in terms of either frequencies or percentages.

Cumulative percentages are particularly useful for describing the relative position of a given score in a distribution. For example, inspection of the cumulative curve in Figure 3.2 shows that a man who scored 33 is quite conservative in comparison to his peers; that is, low scores are associated with traditional views, and only about 20 percent of the group had a score of 33 or less. In contrast, someone who scored 59 could be considered pro-feminist, approximately 90 percent of the men earning this score or below and only about 10 percent a higher score.

Specifying an individual's standing in terms of these cumulative percentage figures often conveys more information than the raw score (how much would it mean to you, for example, simply to be told that on a 60-point examination, you scored 42; is that good or bad?) or even an individual's standing in terms of numerical rank (for example, 27th from the bottom in a class of 64). These cumulative percentages are therefore often used as alternatives to raw scores and are collectively identified as the *percentile* scale.

Using the cumulative percentage curve, we can go back and forth between the two scales. If we start with a raw score and find its equivalent on the percentile scale, we speak of the percentile rank of the score. Thus, *the percentile rank of any given score is a value indicating the percent of cases falling at or below that score.* In Figure 3.2, for example, we can see that the percentile rank of the score 31 is approximately 15. Conversely, we can start with a percentage value and find its score equivalent. This score is called a percentile. Thus, a *percentile is the score at or below which a given percent of the cases fall.* In Figure 3.2 the 60th percentile is approximately 45. (If the distinction between percentiles and percentile ranks seems confusing, don't worry. Just remember that one can go back and forth between raw scores and percentile scores and that percentiles represent the percent who earn the score *or below.*)

The percentile scale, derived from percentages, contains 100 units. A number of points along this scale are given special names. Often encountered are the *quartiles*. The first quartile (known as Q_1) is the special name for the 25th percentile, the second quartile (Q_2) is equal to the 50th percentile, and the third quartile (Q_3) to the 75th percentile. The 50th percentile is also called the *median* of the distribution. Occasionally certain percentiles are referred to as *deciles*: the first decile (or 10th percentile), the second decile (20th percentile), and so forth up to the tenth decile (100th percentile).

The percentile equivalents of raw scores can be determined, we have just seen, by reading them from a graph of cumulative percentages. This method has the drawback of providing only approximate values of percentiles and percentile ranks because of the necessity of estimating values from graphs. Procedures for calculating more precise values will be presented in the next chapter in our discussion of the median.

GENERAL PROCEDURES IN GRAPHING FREQUENCY DISTRIBUTIONS

With brief descriptions of the two types of graphic representations before us, we can now discuss certain procedures applicable to graphing in general.

Selection of Axes and Size of Units

The custom in graphing frequency distributions is to let the horizontal baseline represent scores or measures and the vertical axis frequencies (or percent of cases). This convention should always be followed, since reversal would cause confusion for those used to the customary method.

The physical distance selected to represent a score unit or frequency unit is purely arbitrary, except for convenience, since a graph can be drawn large enough to cover the wall of a room or small enough to require a microscope to be seen. Whatever the size of the graph, however, the general rule for frequency polygons has been to choose the units in such a way that the length of the vertical axis is 60-75 percent of the length of the baseline. This rule may seem picayune but there are reasons for it. Not only does the uniformity produce esthetically pleasing results but there are practical implications as well. The latter become clearer if you examine the polygons in Figure 3.3. Each graph gives a different impression at first glance, even though each represents the same frequency distribution. These differences have been produced by manipulating the physical distances chosen to represent scores and frequency units—or, more accurately, by changing the relationship between the length of the two axes. Since the reader's impression of a frequency distribution can be so easily influenced by such an arbitrary factor, the adoption of an approximately standardized relationship between the two axes facilitates the interpretation of frequency polygons.

Figure 3.3 Three polygons drawn from the same frequency distribution varying in ratio of *X* to *Y* axis.

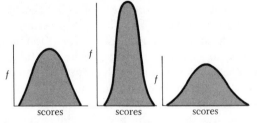

Because the highest frequency appearing in a cumulative frequency curve is greater than that in the polygon of the same data, the same rigid rule does not apply to this type of graphic representation. Most investigators, however, seem to make the baseline equal to or slightly longer than the vertical (cumulative frequency) axis, and it might be well for you to follow this custom whenever convenient.

Labels and Titles of Graphs

Graphs should always be completely labeled. These labels include what each axis represents (such as scores, frequencies, percents) and the numerical values of each. The lowest values for both frequencies and scores should start at the intersection of the two axes, at the lower left of the graph. It is not necessary to indicate or number the position of every possible score along the baseline, but only certain representative ones at regular intervals. Specifically, it is usually sufficient to label class limits (real or apparent) or midpoints of intervals. In frequency polygons, it is most often convenient to indicate the midpoints of intervals, and for cumulative curves the upper real limits. There is no general rule for the frequency scale; as long as enough points are identified to make interpretation easy and these are placed an equal number of units apart, labeling will be considered adequate.

Every graph should also have a concise title containing sufficient information to allow a reader to identify relevant aspects of the data, such as exactly what they measure, their source, and so forth. If several graphs are to be presented, each graph or figure should also be given its own number for further identification (for example, Figure 17). Since a graph without labels and a title is often next to meaningless, you are strongly urged to make "titling" second nature to you, even on statistical exercises.

FORM OF A FREQUENCY DISTRIBUTION

Frequency polygons can occur in an unlimited number of shapes and forms. Since form is often an important bit of information about a distribution, we need to know some of the descriptive terms used to indicate various types.

Some frequency distributions, and consequently the frequency polygons derived from them, are bilaterally symmetrical. That is, if the polygon is folded in half, perpendicular to the baseline, the two sides are identical in shape. A variety of symmetrical distributions is shown in Figure 3.4.

Several distributions that are not symmetrical are shown in Figure 3.5. An asymmetrical distribution in which the scores "tail off" in one direction

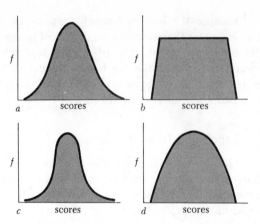

Figure 3.4 Illustrations of symmetrical frequency distributions.

is said to be *skewed*. Such distributions can be more exactly described in terms of degree and direction of skewness. Degree refers to the amount that a distribution departs from symmetry. Thus, distribution *a* in Figure 3.5 is more highly skewed than *c*. Direction of skewness takes account of the fact that in some asymmetrical distributions the measures tend to pile up at the lower end of the scale and to trail off, in terms of frequency, toward the upper end, while in other cases the situation is reversed. Curves *a* and *b* in Figure 3.5 illustrate these two possibilities. Differences in the direction of skewness are indicated by saying that the distributions are positively or negatively skewed. Both as a way of explaining which is positive and which negative and as a device for remembering the difference between the two, the lower or left end of the baseline can be pictured as the negative end (values are low) and the upper or right end as positive (values are high). If the "tail" of a distribution (where scores are relatively infrequent in num-

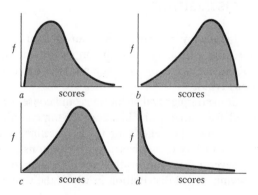

Figure 3.5 Illustrations of skewed frequency distributions.

ber) lies in the negative end of the scale, the distribution is negatively skewed. If the tail extends out toward the upper or positive end, the distribution is positively skewed. Curve *b* in Figure 3.5 is therefore negatively skewed, while *a* is positively skewed.

Certain distinctive forms of distributions that occur with some frequency are given special names. Curve *d* in Figure 3.5 is called a *J curve* (a curve that resembles either a J or its mirror image). This type is found primarily with certain types of socially conditioned behavior, such as a distribution of the number of major crimes committed by members of a large group of individuals. The vast majority have none to their credit, whereas a few members of the group are multiple offenders. In Figure 3.4, *b* is *a rectangular distribution*, the frequencies in every class interval being the same. Distribution *a* in Figure 3.4 approximates *the normal curve*, a specific form of a symmetrical bell-shaped distribution. Many phenomena, including such psychological characteristics as IQ's and certain personality traits, appear to take this form. The normal curve is of such great theoretical importance in statistics that its properties will be discussed in some detail in Chapter 7.

GRAPHIC REPRESENTATIONS OF OTHER TYPES OF DATA

Up to this point we have been considering methods of graphing frequency distributions. We will now discuss very briefly how to represent several other types of data commonly encountered in psychological investigations.

Often in psychological research we try to determine how individuals' performance changes with systematic increases or decreases in the value of some variable. We might ask groups of subjects, for example, to memorize a list of items and then test their recall after one of several time intervals to see the effect of lapse of time on retention. The results of such studies are typically presented in a *line graph*, an example of which is shown in Figure 3.6.[3] The data were obtained from two groups of subjects whose acquisition of a type of conditioned response (increase in magnitude of the galvanic skin response or GSR to a tone in anticipation of an electric shock) was being studied. Subjects were given repeated trials so that the course of learning could be followed. One group avoided the shock on each trial on which their conditioned response to the tone was above a certain magnitude; the other (nonavoidance) group received a shock on a similar number of trials without regard to their response. When the data from the two groups are presented

[3] Adapted from Kimmel, H. D., and Baxter, R. Avoidance conditioning of the GSR. *Journal of Experimental Psychology*, 1964, 68, 482–485.

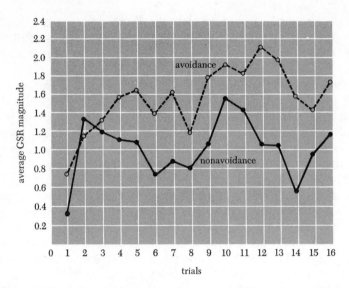

Figure 3.6 A line graph representing the acquisition of a conditioned response in groups tested under avoidance and nonavoidance conditions.

on the same graph, we not only see the changes in response over trials within each group but can easily compare the performance of the two groups. In this instance we see that after the initial trials, the average GSR magnitude of the subjects in the avoidance group is consistently higher than that of the nonavoidance subjects.

In this example a variable along a *continuum* was systematically incremented: individuals were given a number of conditioning trials, with a measure of performance being obtained at each point. Or we might measure the performance of groups of animals whom we deprive of food for varying numbers of hours, or groups of individuals who are high, medium, or low in dogmatism; again, a continuous variable is involved. But sometimes the different groups or individuals whom we measure do not vary with respect to some continuous variable but represent a collection of discrete objects or events that *cannot be ordered*. For example, we might compare on some measure types of animals, different countries of the world, or people of different occupations. The data from such investigations are usually presented in the form of a bar graph to emphasize the groups' discreteness. The illustration found in Figure 3.7 was taken from a study[4] in which subjects were given a number of symbol identification tasks, which they performed under

[4] Data adapted from Howell, W. C., and Kreidler, D. L. Instructional sets and subjective criterion levels in a complex information processing task. *Journal of Experimental Psychology*, 1964, 68, 612–614.

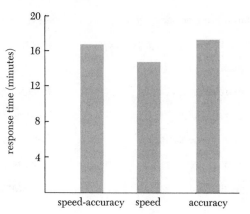

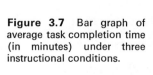

Figure 3.7 Bar graph of average task completion time (in minutes) under three instructional conditions.

instructions emphasizing the importance of speed, of accuracy, or of both. The data in the bar graph show the average time taken to complete the tasks under the three instructional conditions. Since the conditions do not form any sort of ordered continuum, the order in which they appear on the baseline is completely arbitrary.

In graphs of the type shown in the last two illustrations, the response measure (average magnitude of conditioned responses, time to completion) is always indicated on the Y axis. Trials, groups, or conditions are plotted along the baseline. This arrangement is, in a sense, comparable to the one followed in graphing frequency distributions: frequency or percent of the cases is plotted along the Y axis (ordinate) in frequency distributions; frequency or amount of response is plotted also along the Y axis in data involving repeated measures or different groups; scores are plotted along the X axis (abscissa) in frequency distributions and so are groups or trials with other types of data.

A further similarity to graphic representation of frequency distributions is found in the general relationship of the axes; the ordinate should be approximately 60 to 75 percent of the length of the baseline. Needless to say, the requirements for labeling and identifying the graph are just the same with all methods of graphing.

TERMS AND SYMBOLS TO REVIEW

Frequency polygon	Quartile (Q)
Horizontal (X) axis or abscissa	Median (Mdn.)
Vertical (Y) axis or ordinate	Symmetrical distribution

Cumulative frequency (cum f)

Cumulative frequency polygon

Midpoint of class interval

Upper real limit (URL)

Lower real limit (LRL)

Cumulative percentage

Percentile

Percentile rank

Asymmetrical distribution

Positive and negative skew

J-curve

Rectangular distribution

Normal curve

Bell-shaped distribution

Line graph

Bar graph

Measures of
Central
Tendency

4

$\mathbf{W}$e have already seen how group patterns can be made clearer by sorting raw data to form frequency distributions and how to represent such distributions by graphing them. Often, however, we are not interested in group patterns but instead want to characterize a group as a whole. We might have such questions as: How many children does the typical woman college student plan to have? What is the average annual income of college professors? Such questions ask for a single number that will best represent a whole distribution of measurements. This representative number will usually be near the center of a distribution, where the measures tend to be concentrated, rather than at either extreme, where, typically, only a few measures fall. From this fact comes the term "measure of central tendency."

Although a number of measures of central tendency have been devised, we will be studying only three of them: the arithmetic mean ($\overline{X}$), the mode (Mo.), and the median (Mdn.). There is nothing magical about any of these measures of central tendency; each is simply a different method of determining a single representative number. As we shall see, the numbers resulting from each method usually do not agree exactly with each other. The particular measure we compute in any instance depends on which one yields the number that best represents the fact we wish to convey. The specific situations in which each method is most appropriate will be discussed in a later section.

ARITHMETIC MEAN

Definitions of the Mean

The arithmetic mean ($\overline{X}$) is a measure with which most of you are already familiar, popularly known as the "average." The arithmetic mean, or simply "the mean" for short, is the result of the well-known procedure of adding up all the measures and dividing by the number of measures. Defining the term more formally, *the mean is equal to the sum of the measures divided by their number.* A much more compact way to express this verbal definition is by using symbols, a kind of mathematical shorthand, as shown in the formula below:

$$\overline{X} = \frac{\sum X}{N} \qquad\qquad (4.1)$$

where: $\overline{X}$ = arithmetic mean
$\sum$ = Greek capital letter sigma, meaning the "sum of" a series of measures
X = a raw score in a series of measures
$\sum X$ = the sum of all the measures
N = number of measures

Before proceeding with a discussion of the mean ($\overline{X}$), we will take time out to explain in more detail the meaning of the Greek capital letter sigma (Σ), one of the most frequently used symbols in statistics. The presence of Σ in any formula indicates that a group of numbers is to be added or summed. Since for convenience we usually refer to any raw score in a frequency distribution as X, Σ X indicates the total obtained by adding together all the scores belonging to a distribution. Sometimes we deal with two frequency distributions (for example, scores made by a group of business school and social science majors on a measure of political conservatism). To distinguish between the two distributions we might call one set X's (let us say, business majors) and the other Y's (the social science majors). If we wrote the symbols Σ Y, we would mean the sum of all the scores for the social science majors (the Y's). In like manner, Σf stands for the sum of all the frequencies in a distribution. Or we might arbitrarily choose the symbol A, and then Σ A is the sum of all of the A scores, whatever they happened to be, and so forth. The expression Σ X/N, then, is shorthand for the sum of all the raw scores (X's), divided by N (their number). The number that results from these operations is the mean of the distribution of X's, symbolized as $\overline{X}$ (pronounced "X-bar"). In parallel fashion, the mean of a distribution of Y's is symbolized as $\overline{Y}$, of A's as $\overline{A}$, and so on.

Another definition or characteristic of the mean ($\overline{X}$) is important for us to state, since it contributes to an understanding both of $\overline{X}$ and of the concept of variability, a topic to be considered in Chapter 5. *The mean may be described as that point in a distribution of scores at which the algebraic sum of the deviations from it (the sum of the differences of each score from $\overline{X}$) is zero.* Or, expressed differently, $\overline{X}$ is a kind of "point of balance," where the sum of the deviations of the scores above $\overline{X}$ is equal in absolute value (without regard to plus or minus sign) to the sum of the deviations below $\overline{X}$. In order to clarify these statements, we will first introduce a kind of analogy. Imagine that the horizontal line shown in Figure 4.1 is an old-fashioned teeter-totter. Suppose further that each square drawn in Figure 4.1 represents a child, each of equal weight, seated at the place indicated. We can see that the fulcrum has been placed at a point where the board balances. But why does the board balance at this particular point? First observe that we can give each child a "value" in terms of his distance from the fulcrum (-1, -2, -3, $+1$, $+2$, $+3$). These distances we will call deviations from the fulcrum. The absolute sum (that is, ignoring signs) of the deviations of the

Figure 4.1 Representation of a balanced teeter-totter with "children" of equal weight placed at the distance indicated.

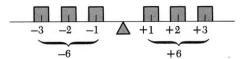

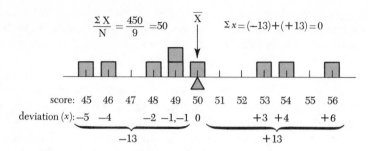

$$\frac{\Sigma X}{N} = \frac{450}{9} = 50 \qquad \overline{X} \qquad \Sigma x = (-13) + (+13) = 0$$

score: 45 46 47 48 49 50 51 52 53 54 55 56

deviation (x): −5 −4   −2 −1,−1 0   +3 +4   +6

−13   +13

Figure 4.2 Distribution of raw scores and deviations of these scores from $\overline{X}$, showing that the algebraic sum of the deviations is zero.

three children to the right of the fulcrum is the same as the absolute sum of the deviations of the children to its left, namely, 6 units. From painful experience, you know that the balance of the teeter-totter would be destroyed if a child were moved from one position to another, even on the same side (for example, from −1 to −3). When we disturb the values of the deviations so that their sums are not equal on both sides of the fulcrum, the balance is destroyed and the fulcrum must be moved to restore it.

Let us now relate this analogy to $\overline{X}$, which we have already said is a kind of point of balance. First, let us find $\overline{X}$ by the method of adding up the scores and dividing by N, as in the example given in Figure 4.2. Since there are 9 scores in Figure 4.2 and their sum is 450, $\overline{X}$ is 50. Next, let us find the deviation or distance each score is from $\overline{X}$. Each deviation is designated by the symbol x and is obtained by subtracting $\overline{X}$ from each score ($x = X - \overline{X}$). Obviously, scores with values greater than $\overline{X}$ will be plus deviations, those with lesser values, minus deviations. If we now add the plus deviations (those above $\overline{X}$) and the minus deviations (those below it), as was done in Figure 4.2, we find that the absolute value of the two sums is equal; retaining signs, the two sums cancel each other so that the sum of the *total* deviations is zero ($\Sigma x = 0$). Take as another example the numbers 5, 9, and 16. $\overline{X}$ of these numbers is 10, the sum of deviations below it is -6 $[(-5) + (-1)]$ and above it $+6$; the algebraic sum of the deviations equals 0. Thus, in any distribution of scores, $\overline{X}$ is a "point of balance," comparable to the fulcrum in the teeter-totter example, in that the algebraic sum of the deviations of the scores above and below it (Σx) always equals zero.

Methods of Computation

The basic formula that defines a statistical concept, such as that for $\overline{X}$, is usually relatively simple in appearance, but actually applying it to data often turns out to be long and tedious. To minimize effort, so-called "computational formulas" are usually derived from the basic one. These often look more difficult, as they involve more symbols, but are time-savers when any statistic is actually to be computed. In the following sections several computational formulas for $\overline{X}$ will be explained, in addition to the application of the basic one.

Computing $\overline{X}$ from ungrouped data. When $\overline{X}$ of ungrouped data is to be calculated, the basic definitional formula should be followed: the scores added and the result divided by the number of scores. An application of the basic formula is demonstrated in Table 4.1. This process is simple and should be followed if only a few scores are involved. If the number of measures is large, however, it is prohibitively time-consuming. The use of frequency distributions, already found to be convenient for other purposes, will reduce our labor considerably in such instances.

Table 4.1 Calculation of $\overline{X}$ from ungrouped data.

X
13
13
11
16
18
13
19
15

$$\sum X = 118$$

$$\overline{X} = \frac{\sum X}{N} = \frac{118}{8} = 14.75$$

Computing $\overline{X}$ from a frequency distribution. Suppose that 70 children were given a perceptual judgment task to determine their susceptibility to optical illusion. The simple frequency distribution in Table 4.2 gives the number of errors these children made on the task. Note that most of the scores occur more than once. For example, two children made 9 errors and eight made 14. We could get $\sum X$ by summing the 70 individual scores, duplicates and all (for example, adding two 9's, eight 14's). However, we could save time if we first multiplied each score by the number of individuals receiving it. Thus, 8×14 will result in the same answer as adding 14 eight times and give it to us quickly. If we then add all of the products (each the result, you remember, of multiplying every X by its corresponding frequency, f), we will have the sum of all the X's with far less labor than if we had summed each raw score separately. The remaining step needed to obtain $\overline{X}$ is the usual one: dividing the sum of the scores by N.

We can express these operations symbolically by amending the basic formula for $\overline{X}$ to take account of the fact that we are now dealing with a frequency distribution. The amended formula is reproduced below and its application demonstrated in Table 4.2.

$$\overline{X} = \frac{\sum (fX)}{N} \tag{4.2}$$

where: $\sum$ = the sum of the quantity that follows (here, all the fX's)
  X = the midpoint of a class interval
  (fX) = a midpoint multiplied by its corresponding frequency (f)—that is, by the number of cases within a class interval
  N = total number of cases, equal to the sum of the frequencies $(\sum f)$

Table 4.2 Calculation of $\overline{X}$ from a simple frequency distribution, where $i = 1$.

X	f	fX	X	f	fX
28	2	56	16	4	64
27	1	27	15	5	75
26	1	26	14	8	112
25	0	0	13	6	78
24	0	0	12	5	60
23	3	69	11	3	33
22	1	22	10	3	30
21	2	42	9	2	18
20	3	60	8	3	24
19	5	95	7	1	7
18	6	108	6	0	0
17	4	68	5	2	10
				$\sum f = 70$	$\sum fX = 1084$

$$\overline{X} = \frac{\sum fX}{N} = \frac{1084}{70} = 15.49$$

If necessary, the formula can also be applied to a *grouped* frequency distribution, one in which the width of the class interval, i, is greater than one. This has been done in Table 4.3 for the data utilized in the previous table, now grouped into class intervals with a width of three. In applying the formula to grouped distributions, it should first be noted that the symbol

X is now the *midpoint* of a class interval. Actually, X in the formula for $\overline{X}$ of a frequency distribution is always a midpoint, even when $i = 1$ as in Table 4.2. In the latter instance we can think of a series of class intervals, each with a width of 1 (for example, class 5, 6, and 7). Thus, each raw score, X, is in a sense its own midpoint.

Table 4.3 Calculation of $\overline{X}$ from a grouped frequency distribution, where $i = 3$.

Class Interval	X	f	fX
26–28	27	4	108
23–25	24	3	72
20–22	21	6	126
17–19	18	15	270
14–16	15	17	255
11–13	12	14	168
8–10	9	8	72
5–7	6	3	18
		$\sum f = 70$	$\sum fX = 1089$

$$\overline{X} = \frac{\sum fX}{N} = \frac{1089}{70} = 15.56$$

SOURCE: The data are derived from Table 4.2.

Going back to Table 4.3, multiplication of each midpoint by its corresponding frequency (fX), as done in the last column of the table, gives the (approximate) total of the scores in each interval. By adding all of the entries in the fX column, we get $\sum (fX)$. This is the (approximate) equivalent of the $\sum X$ that would have been obtained if ungrouped scores had been added together. $\overline{X}$ is determined, of course, simply by dividing $\sum (fX)$ by the total number of *cases*, N (*not* the number of class intervals).

While the foregoing method is convenient, the use of grouped frequency distributions introduces a source of error into the computation of $\overline{X}$. Underlying its use is the assumption that all scores in a particular interval fall at the midpoint of the interval or, more accurately, that $\overline{X}$ of the scores in a specific interval is the same as the midpoint of the interval. If this is true, we can get the total of the scores in the interval by multiplying the number of measures (f) by the midpoint (X).[1] This assumption rarely is fulfilled exactly.

---

[1] Since $\overline{X} = \sum X/N$, by rearrangement of terms, $\sum X = \overline{X}(N)$. Thus, if $\overline{X}$ of the scores in a particular interval is equal to the value of the midpoint, we can find the total of the scores in the interval ($\sum X$) by multiplying the midpoint by the frequency [$\sum X = \overline{X}(N)$].

The net effect of such errors on $\overline{X}$ of the total distribution is not great, however, since in the intervals below $\overline{X}$ the errors tend to be in the direction of underestimation (scores usually pile up at the upper end of the interval, thus making the true $\overline{X}$ of the scores in the interval higher than the midpoint), and in intervals above $\overline{X}$ the opposite error more frequently occurs (scores tend to concentrate at the lower part of the interval). Thus, when all intervals are considered together, the errors in one direction tend to cancel the errors in the other, so that $\overline{X}$ for the total distribution is fairly close to the more accurate figure that would have been obtained had ungrouped data been used (in our example, 15.49 vs. 15.56).

MODE

The mode (Mo.) may be either computed or estimated roughly from inspection of the data. Estimation by inspection is the only method we will consider here.

In ungrouped data, the mode is defined as that score which occurs most frequently. For data grouped into class intervals Mo. is the midpoint of the interval with the greatest frequency. Thus, in Table 4.1, showing ungrouped data, Mo. is 13, since this score occurs three times and all others only once. In Table 4.3 Mo. is 15, this score being the midpoint of the interval with the greatest frequency. The value of Mo. may change, even with the same set of data, as the width of the class interval changes. Compare, for example, the Mo. of Tables 4.2 and 4.3.

Sometimes a frequency distribution, like a camel, will have not one hump but two, indicating two points of maximum frequency rather than one. (On rare occasions there may be even more than two.) This type of curve is called *bimodal* in contrast to the more usual *unimodal* or single peaked curve. Both humps are identified as modes.

MEDIAN

We have already encountered the median (Mdn.) in our discussion of percentiles, since the median is just another label for the fiftieth percentile. Thus, *the median may be defined as that score point at or above which 50 percent of the cases fall and at or below which 50 percent of the cases fall.* In some populations (that is, groups consisting of all individuals who exhibit the specified characteristic, as opposed to a sample of individuals), the distribution may be a peculiar one in which this definition may be satisfied by more than one point (for example, by all the values from X = 4.0 to

X = 5.0). In such an instance we would have to say that the median is the *set of points* satisfying the condition stated in our definition. We will restrict ourselves here to discussing computational procedures for sample distributions yielding only one point.

Computation of the Median and Other Percentiles

Computation from Frequency Distributions. Procedures for computing the median or 50th percentile of a distribution are the same as for computing any other percentile and can be explained by referring to the grouped frequency distribution in Table 4.4. Our first step is to determine how many of the total number of cases constitute the given percent. In finding the median of the data in Table 4.4, we must therefore take 50 percent of N, which gives us 30 cases (60 × .50). The score point we want, then, is 30th from the bottom of the distribution. Going up the *cum. f* column we find that the 30th case falls someplace in the interval 40–44, since up to 39.5, the lower real limit of this interval, there is a total of 22 cases (too few) and up to 44.5, the upper real limit of the interval, 33 cases (too many). We need to go up through the 8th case of the 11 in the interval 40–44 to make up the needed 30 (22 + 8 = 30). The raw-score value of this 8th case will give us the median.

Now we encounter a problem. The score of the 8th case cannot be determined because we do not know the placement or distribution of the 11

Table 4.4 Grouped frequency distribution used in demonstrating computation of percentiles and percentile ranks.

Class Interval	f	cum. f
65–69	1	60
60–64	2	59
55–59	5	57
50–54	7	52
45–49	12	45
40–44	11	33
35–39	8	22
30–34	7	14
25–29	4	7
20–24	3	3
N = 60		

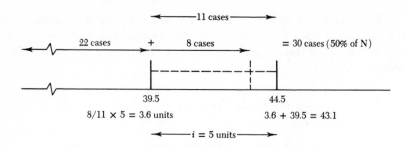

Figure 4.3 Graphic representation of obtaining a percentile.

cases in the interval. We can, however, *estimate* its value by assuming that the cases in the interval are evenly distributed; that is, that the same number of cases fall at each score point, so that the shape of the intrainterval distribution is rectangular. Let us now examine Figure 4.3 to see the implications of this assumption. If the distribution is rectangular, then the 8th case in the interval ends $8/11$ of the way up the interval. If we take $8/11$ of the score units in the interval (that is, $8/11$ of i), we find the "score distance" the upper end of the 8th case is from the lower real limit. This distance ($8/11$ of 5) turns out to be 3.6 score units. Adding 3.6 to 39.5, the lower real limit, gives us 43.1. This score, which is the 30th from the bottom of the complete distribution, is the 50th percentile or the median.

The procedures we have just discussed for obtaining the median can be described by a sort of "cookbook" recipe, which can be followed step by step and used to determine any percentile.

A. Translate the percent of cases given in the percentile into number of cases (the nth case).

B. Find the score corresponding to the upper end of the nth case:
(1) Locate by inspection of the *cum. f* column the class interval in which the nth case lies. This is the interval containing the percentile.
(2) Subtract from the nth case the *cum. f* of the interval *below* the one in which the nth case is contained. This tells us the number of cases we need from the interval.
(3) Divide the value found in step 2 by the number of cases (f) in the interval containing the given percentile.
(4) Multiply the quotient obtained in step 3 by i.
(5) Add the number from step 4 to the lower real limit of the class in which the percentile (the nth case) is contained. This score is equal to the desired percentile.

These steps can be summarized in the following formula.

$$\text{Score} = \text{lower real limit of int.} + i \left(\frac{[n\text{th case}] - [cum. \, f \text{ in int. below}]}{\text{no. of cases in int.}} \right) \quad (4.3)$$

where: nth case = the number of cases corresponding to given percent (for example, for 37th percentile when $N = 40$, nth case = 37 percent of 40)

int. = class interval in which the nth case is contained (determined by inspection of *cum. f* column)

int. below = class interval immediately below the one in which the nth case is contained

We can demonstrate the use of this method by applying it to another concrete example. Let us find the 80th percentile (or 8th decile) of the distribution in Table 4.4.

A. nth case = 80 percent of N or 48, where N = 60.

B. Find the score corresponding to the nth case:

$$\text{lower real limit of int.} + i \left(\frac{[n\text{th case}] - [cum. \, f \text{ in int. below}]}{\text{no. of cases in int.}} \right)$$

thus: $49.5 + (5)([48 - 45]/7) = 51.6$.

This same method of computing percentiles can be used with simple frequency distributions with one slight modification: each score value is treated as *the midpoint of a class interval*. If the score values of a simple frequency distribution consisted of integers ranging from 123 upward, for example, 123 would be regarded as the midpoint of the interval whose real limits were 122.5–123.5, 124 the midpoint of the interval whose real limits were 123.5–124.5, and so on.

We have not discussed the strange case in which there is a zero frequency in the interval in which a particular percentile, such as the median, falls. This means that the median or other percentile has several values rather than the single value promised in our definition of percentiles. For these unusual cases with zero in the relevant interval, we recommend defining the median or other desired percentile as the *midpoint* of the interval with zero frequency.

Computation from Ungrouped Data. An alternate definition of the median that is useful with ungrouped data specifies that the median is the *middle score value* in a set of measures. (A parallel definition may be offered for any percentile by substituting the appropriate score position.) With this method, scores are first ranked in order of magnitude. If the number of cases is *odd*, the position of the median is equal to $(N + 1)/2$. We count up from the bottom until we reach the score that has the $(N + 1)/2$th position. In the set of five numbers, 9, 11, 15, 16, and 18, $(N + 1)/2$ is equal to 3; the third case from the bottom is 15; this is the middle number or the median. If the number of cases is *even*, we count up to the cases that have the positions $N/2$ and $(N + 2)/2$ and then find the *midpoint* of these two scores. In the ordered set of six numbers, 3, 5, 6, 8, 12, and 13, we therefore count up to the third and fourth cases, finding the scores 6 and 8. The midpoint of these two scores is 7, the median of the distribution.

The median for ungrouped data may take a slightly different value from the one that would be found if we applied the computational method for a frequency distribution that we discussed earlier. This occurs because of the treatment in the latter approach of each score as an interval (for example, 9.5–10.5), rather than simply as an exact value (for example, 10.000 . . .).

Computation of Percentile Ranks. We often want to report individuals' scores in terms of percentile rank, rather than raw scores since the latter are seldom meaningful in and of themselves. Suppose, for example, we wanted to know the percentile rank of the score 47, found in the distribution in Table 4.4. What has to be determined is the *number* of cases receiving a score of 47 or less, and from there, the *percent* of cases. This percent gives us the percentile rank.

Looking at Table 4.4, we discover that 47 is in the interval 45–49 and, from inspecting the *cum. f* column, that 33 cases fall below the lower real limit of the interval in which our score of 47 is contained. To obtain the number of cases scoring 47 or below, we have to add to the *cum. f* of 33 the number of cases lying between the lower real limit, 44.5, and our score, 47. By assuming that the 12 cases in the interval are rectangularly distributed, just as we did earlier in computing percentiles, we can determine this number quite simply. The width (i) of the interval is 5 score units and our score lies 2.5 units up from the lower real limit $(47 - 44.5 = 2.5)$. We therefore estimate that $^{2.5}/_5$ or $^1/_2$ of the 12 cases lie between our score and the lower real limit. We now add $^1/_2$ of 12, or 6, to the *cum. f* up to the interval, which gives us $6 + 33$ or 39 cases. Now the going is easy, since all we have to do is to translate number of cases into the *percent* of N. This we do by dividing our number by N and multiplying the result by 100, which gives us $^{39}/_{60} \times 100$. Hence the answer to our problem is that the score of 47 has a percentile rank of 65.

These procedures for obtaining percentile ranks from frequency distributions can be described by a step-by-step "recipe."

A. Determine the number of cases falling below the given score:
 (1) Subtract from the score the lower real limit of the interval in which it is contained.

(2) Divide the result of step 1 by the width of the interval (i).
(3) Multiply this quotient by the number of cases (f) in the interval.
(4) Add the product obtained in step 3 to the *cum. f* of the interval *below* the one in which the score falls.

B. Translate the number of cases below the score into percent of cases:
(5) Divide the result of step 4 by N and multiply the product by 100. This gives us the percentile rank of the score.

These steps can be summarized in the following formulas, which should be easy to apply after a few practice problems have made the terms in them familiar.

(1) No. of cases = *cum. f* in int. below

$$+ \left(\frac{\text{given score} - \text{lower real limit of int.}}{i} \right) (f \text{ in int.}) \qquad \textbf{(4.4)}$$

(2) Percentile rank $= 100 \left(\dfrac{\text{no. of cases}}{N} \right)$

where: int. = class interval which contains the given score whose percentile rank is being found
        int. below = class interval immediately below the one which contains the given score

To illustrate the use of these formulas we will find the percentile rank of the score 28 in Table 4.4.

(1) Number of cases $= 3 + ([28 - 24.5]/5)(4) = 5.8$
(2) Percentile rank $= 100(5.8/60) = 9.67$

COMPARISON OF THE MEAN, MEDIAN, AND MODE

In any unimodal symmetrical distribution the values of $\overline{X}$, Mdn., and Mo. are the same. This is true because the same point on the baseline occurs most frequently (Mo.), divides the number of cases into the upper and lower 50 percent (Mdn.), and is the point of balance ($\overline{X}$). As a curve departs from symmetry and becomes skewed, however, the values of the three measures of central tendency show differences. The relative positions of the three measures, however, will usually be the same for all single-peaked non-symmetrical distributions, as shown in Figure 4.4. Mo. in any skewed

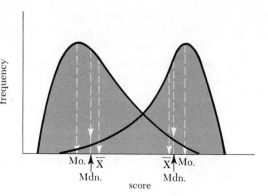

Figure 4.4 Relationship between X̄, Mdn., and Mo. in skewed distributions.

distribution occurs, of course, at the highest point of the curve. X̄ falls someplace toward the "tail" of the distribution. This can be understood if you remember that X̄ is a "point of balance" and is therefore very sensitive to the extreme deviations contained in the tail. Mdn. in a skewed distribution lies between the X̄ and Mo. Thus, the order of the three measures, when we start at the hump and proceed toward the tail, is (1) Mo., (2) Mdn., and (3) X̄.

Because there is this relationship among the values of the three measures, we ordinarily can tell whether a distribution is skewed, and if so, the direction, merely by being told the values of any two measures of central tendency. If we read that X̄ of a distribution was 26 and that Mdn. was also 26, we would infer that the distribution was symmetrical, since only in this case do the measures coincide. But if we were told that X̄ was 109 and Mo. from the same distribution 102, we would know that the distribution was positively skewed. That is, since X̄ always lies closest to the "tail" in a skewed distribution and in this case is greater than Mo., the tail must be in the upper, positive end of the scale. The value of Mdn. in this example could be roughly estimated—greater than 102 and less than 109—since Mdn. always falls between X̄ and Mo.

When to Use the X̄, Mdn., or Mo.

In deciding which measure of central tendency to compute, we must consider a number of factors, including how much time is available, the characteristics of the data involved, and the purpose for which the measure is intended. Some of these considerations are discussed below.

Stability of Measures. The three measures of central tendency differ with respect to their consistency or stability from sample to sample. That is, if we tested a number of successive groups (or samples) of individuals drawn from some larger group, the values of $\overline{X}$, as well as those of Mdn. and Mo., would vary from sample to sample. For example, we might test several samples of middle-aged businessmen to determine their cholesterol level. We would find that the mean was not the same for each group, nor was the Mdn. or Mo. But we would discover that $\overline{X}$ showed less variability from sample to sample than the other two (that is, was most stable) and that Mo. was the least consistent. That is, if we took a large number of samples from a large group and computed $\overline{X}$, Mdn., and Mo. for each, we would find that the values of the $\overline{X}$'s differed less among themselves than the Mdn.'s and Mo.'s, whereas the Mo.'s varied the most. Since we frequently test small groups or samples in order to estimate the characteristics of the larger group or population to which the samples belong, stability is a desired virtue for a statistical measure. With respect to consistency, then, it should be remembered $\overline{X}$ is most satisfactory, Mo. is least satisfactory.

Subsequent Manipulations of the Data. In later chapters we will discover that $\overline{X}$ may be used in further statistical operations, yielding additional kinds of information about data. In contrast, once Mo. and Mdn. have been computed, little more can be done with them, especially Mo. Since information beyond a measure of central tendency usually is to be obtained from data, computation of $\overline{X}$ becomes almost obligatory.

Time Factors. A trivial but often practical consideration is the time taken to compute each measure. If a measure of central tendency is needed in a hurry, Mo. can be rapidly obtained by inspection and would therefore be preferred over the other measures.

Characteristics of the Data. 1. *Skewed distributions.* When a distribution is markedly nonsymmetrical, it is possible to give a distorted impression of the data when reporting central tendency. Take, for example, a frequency distribution of incomes in the United States. Such a distribution is extremely skewed in a positive direction, incomes trailing off, in terms of frequency, from the modest modal point to fabulous incomes of several million dollars each. The modal and median incomes will be much lower sums than $\overline{X}$, which is drawn far out toward the tail. It might be good capitalistic propaganda to give $\overline{X}$ as the "average income in the U.S.," but it would give an uninformed reader, who is likely to assume that this figure represents the most frequent income, or the income of the common man, a very inflated view of our personal wealth. A Marxist, on the other hand, would be more likely to select Mo. to represent us, since this measure would have the smallest value. Neither way is completely honest, of course, and what is usually done in the case of extremely skewed distributions is to report all three measures. From these the reader can infer the direction and amount of skewness and properly interpret the data.

2. *Some special cases.* In some instances, unusual characteristics of the data, other than skewness, dictate the measure of central tendency to use. A peculiarity that sometimes occurs is that one extreme of a distribution is not available for testing. For example, if we gave an intelligence test to a sample of school children in order to estimate the central tendency of IQ's of *all* children of similar age, the lower extremes (mentally retarded) would not be represented in our sample, as these children are not in school.

In such a situation, computation of $\overline{X}$ would yield a very distorted estimate of $\overline{X}$ for children in general, since this measure is highly sensitive to extreme deviations and these are missing at the lower end. Mo. or Mdn. would be more appropriate, since each would give a closer estimate of the value of central tendency that would have been obtained if the lower extremes had been available.

Occasionally, representative cases are available for testing, but exact scores cannot be obtained from cases falling in one or both extremes of the distribution. In learning experiments involving animals, for example, running-time from one end of a path to the other may be used as a response measure. Often a few animals refuse to run and after sitting at the starting point for, say, five minutes are removed from the apparatus. Their exact running-times are therefore unknown, since they might have run had they been left a second longer or they might have stayed there forever. Since $\overline{X}$ requires the exact value of every measure, it should not be computed. Mdn. could be used, however, since an approximate score (five minutes plus) for the extreme cases is sufficient.

Many further examples of very special situations in which one measure is appropriate (or inappropriate) could be given. Instead, you will merely be reminded that every set of data should be examined to determine whether any specific problems exist and which measure of central tendency would be most appropriate.

Specific Purpose for Which Measure is Intended. Sometimes a measure of central tendency is intended for some special use rather than as a purely scientific description of data. Such a purpose will often determine the measure to be employed, especially when a distribution is skewed so that the values of the $\overline{X}$, Mdn., and Mo. disagree markedly. Occasionally, the "typical case" is wanted, thus calling for Mo. A furniture manufacturer, hoping to make money on volume of sales, might want to know the size of the typical living room, the modal dimensions, so that he could design furniture scaled to suit the largest single consumer market. At other times, the "middle case" is desired, the median individual. Test grades are sometimes given on this basis, Mdn. being the dividing line between B's and C's.

In summary, $\overline{X}$ is the most generally preferable measure of central tendency, particularly in a nearly symmetrical distribution, since it has the greatest stability and lends itself to further statistical manipulations. Mdn. is the middle case and is usually considered most appropriate when the distribution shows peculiarities: marked skewness, missing cases, and the like. Finally, Mo. is utilized in situations in which a quick, rough estimate of central tendency is sufficient or the typical case is wanted. Although these are good general rules, remember that they are not exhaustive and that each set of data should be examined to see exactly which measure or measures are best in a specific situation or for a particular need.

TERMS AND SYMBOLS TO REVIEW

Measure of central tendency	Point of balance
Arithmetic mean ($\overline{X}$)	Deviation score (x or $X - \overline{X}$)
Mode (Mo.)	Percentile
Median (Mdn.)	Percentile rank
X	Stability
Sum of (Σ)	Sample

Variability

5

THE NEED FOR AN INDEX OF VARIABILITY

Consider the three polygons in Figure 5.1. They are all unimodal symmetrical distributions, are based on the same number of cases, and have the same mean—yet they are quite different. What do these differences imply and how might they be described? Rather than answering these questions in the abstract, let us suggest some examples of what the distributions might represent.

Assume that three friends, Al, Ben, and Carl, spent many hours together on a golf driving range. They decided to keep a record of the distance they each hit 500 balls. Markers on the range allowed them to estimate their drives to the nearest 20 yards. Grouping the 500 drives into class intervals gave the three distributions as shown in Figure 5.1. Looking at the distributions, we see that Al is quite inconsistent; he sometimes gets off a drive that a pro would envy but sometimes he barely rolls the ball off the tee. Carl, on the other hand, is more predictable; he rarely dubs a drive badly but he never gets the tremendous distance that Al sometimes does. Ben is in between; most of his shots are very similar, but occasionally he goes wild, with results that are either very good or disastrous.

Let us assume instead that the distributions represent the scholastic

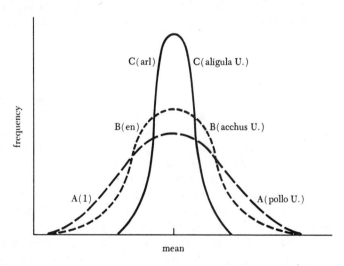

Figure 5.1 Three hypothetical distributions differing in variability.

aptitude scores of freshmen at three different universities. At Apollo University some of the students look like budding geniuses while others will probably have a tough time making it through the first semester. At Caligula University there are no such extremes; as a group the freshmen are much more alike in aptitude than those at Apollo. Bacchus University is in between; occasional freshmen are their professors' delight or despair, but most of them have a respectable, quite unexceptional degree of aptitude.

What terms might be used to describe the differences among the trio of distributions in Figure 5.1? We might say that the members of distribution C (Carl, Caligula U.) are more *homogeneous* than those in the other two distributions. Conversely, members of distribution A (Al, Apollo U.) are more *heterogeneous* than those in the other distributions. We might also say that distribution A has a greater spread than the others—a greater scatter or dispersion or, most commonly, greater *variability*. Distribution C exhibits the least variability among the measures, and B an intermediate amount. In the first illustration (the three golfers) the variability is *within* an individual, so that the curves may be said to represent differences in *intraindividual variability*. That is, repeated measures of the same activity were taken on the same person and each curve indicates how much that person varied from occasion to occasion on the same task. In the case of the second set of distributions (scholastic aptitude) the values represent *interindividual* variability—differences in aptitude among individuals at a particular college.

These examples illustrate that variability is an important characteristic of distributions. If we are to describe distributions adequately, we must include an expression that reflects the amount of variability. We shall discuss two statistics devised for this purpose, the *range* and the *standard deviation*.

RANGE (R)

As we learned in Chapter 2, *the range* (R) *is the "distance" from the lowest to the highest score in a distribution*. Or, highest score minus lowest score is R. If the lowest aptitude score in Bacchus U. is 400 and the highest 600, R is 200 (600 − 400 = 200). If the shortest drive in Al's 500 shots is 40 yards and his longest 290, his golf range is 250 yards.

As a measure of variability R has much the same status as Mo. does as a measure of central tendency. It is useful only as a gross descriptive statistic, although, as we shall see later, it can form the basis for a quick and easy estimate of another statistic that requires considerable computation.

The most obvious difficulty with R as a measure of variability is that its value is wholly dependent upon the two extreme scores. This causes several problems. First, one or both of the extreme scores may be capricious, so that R will be equally capricious. This becomes quite evident when we are working with a small sample of measures, as we often do in psychological research. Suppose we had ten measures as follows: 11, 14, 14, 16, 19, 20, 21, 24, 26, 42. R is 42 − 11, or 31. Now if we eliminate the tenth case, R becomes 26 − 11, or 15. The elimination of one score halved the range. Of course, any measure of variability must represent changes in variability when these occur, but an ideal measure would not fluctuate so much as R does when a single score is eliminated as in the case above.

Even if the two extreme scores are representative, there is another difficulty with R, which we may illustrate by comparing distributions A and C in Figure 5.1. Both distributions have the same R, but the bulk of the scores in distribution C cluster more closely together than those in distribution A. The R, besides being unstable, is insensitive to the shape of the distribution of the scores between the two extremes. For these reasons R is thought of as only a rough index of variability. The statistic that is almost universally used as a measure of variability, the standard deviation, overcomes these problems by being responsive to the exact value of every score in a distribution.

STANDARD DEVIATION (SD)

The standard deviation (SD) requires several computational steps and is a statistical abstraction whose meaning is not easy to grasp. We therefore shall work into it gradually.

In Chapter 4 we discussed the deviation of raw scores from $\overline{X}$. For each raw score we can calculate a deviation score by subtracting the mean from the score, $(X - \overline{X})$. Each of these deviations, you recall, is symbolized by x, so that $X - \overline{X} = x$. If all of the scores are closely clustered around the mean (if the distribution is quite homogeneous), then the raw scores deviate very little from the mean and the x's therefore are low in value. In a heterogeneous distribution in which scores are widely dispersed, on the other hand, the x's are more variable in value, since a number of scores lie far from the mean. Deviation scores, then, reflect the *variability* among a collection of raw scores. How might we capitalize on this fact to get a single number that will allow us to describe the amount of variability in a particular distribution?

One method that might occur to you is to get the mean of these deviation scores. That is, add up all the deviations and divide this sum by the number of cases ($\sum x/N$). This is an excellent beginning, but the procedure suffers from a lethal embarrassment. The sum of the deviations from the mean, we discovered in Chapter 4, is *always* equal to zero (that is, the sums of the deviations above and below the mean are equal except for sign). Thus the mean of the deviations, $\sum x/N$, will also equal zero—hardly an informative figure. We could, however, disregard the signs of the deviations and treat the x's as absolute numbers (for example, for $\overline{X} = 10$, the x's of the raw scores

13 and 7 would each be treated as 3 rather than $+3$ and -3). We could then go ahead and compute the mean of these absolute x's. The resulting statistic has come to be known as the *average deviation*.

Although the average deviation provides a satisfactory number, for technical reasons it is rarely used as a measure of variability. However, a few steps beyond the average deviation brings us to a statistic that is almost universally used as a measure of variability, the *standard deviation* or SD.

In obtaining SD, we avoid the problem of having the plus and minus x's cancel each other by squaring each x so that all the resulting numbers (x^2's) are positive. These squared deviations are summed and then divided by N, thus giving us the *mean* of the *squared deviations*. Finally, to get the SD, we extract the *square root* of this mean. Summarizing these steps, we can describe the SD as the square root of the mean of the squared deviations. To put the procedure into symbolic form, we have:

$$SD = \sqrt{\frac{\sum x^2}{N}} = \sqrt{\frac{\sum (X - \bar{X})^2}{N}} \tag{5.1}$$

Before extracting the square root of $\sum x^2/N$, we have a quantity called the *variance*. The variance has a great deal of use in more advanced statistical techniques and will be discussed in more detail in Chapter 12. For the moment, simply remember that when we refer to the variance, which can also be identified as SD^2, we refer to the quantity $\sum x^2/N$. The *square root* of this number, as we have said, is the standard deviation or SD.

It may be useful at this point to run through a complete computational procedure for obtaining the variance and the standard deviation. Table 5.1 shows, on the left, the scores of eight 4-year-old girls on a test of manual dexterity and, on the right, the scores of eight 4-year-old boys on the same test. The basic computational steps, which are shown in the table for each distribution, can be summarized as follows:

(1) Find $\bar{X}$ by summing the raw scores and dividing by N.
(2) Subtract $\bar{X}$ from each raw score to obtain each x. (The algebraic sum of these x's must equal zero, or there is an error in calculating $\bar{X}$ or in subtracting.)
(3) Square each x value.
(4) Sum the x^2's and then divide by N. (This quantity is SD^2 or the variance.)
(5) Obtain the square root of the result to get SD.

Before leaving Table 5.1, let us go over the results of our calculations. You will observe that although the means of the boys and the girls are the

Table 5.1   Basic steps in calculating a standard deviation.

	Girls			Boys	
X	$x = X - \bar{X}$	x^2	X	$x = X - \bar{X}$	x^2
15	4.5	20.25	18	7.5	56.25
14	3.5	12.25	16	5.5	30.25
12	1.5	2.25	14	3.5	12.25
10	−.5	.25	13	2.5	6.25
10	−.5	.25	9	−1.5	2.25
9	−1.5	2.25	6	−4.5	20.25
8	−2.5	6.25	5	−5.5	30.25
6	−4.5	20.25	3	−7.5	56.25
$\sum X = 84$	$\sum x = 0.0$	$\sum x^2 = 64.00$	$\sum X = 84$	$\sum x = 0.0$	$\sum x^2 = 214.00$

$$\bar{X}_G = \frac{\sum X}{N} = \frac{84}{8} = 10.5 \qquad\qquad \bar{X}_B = \frac{84}{8} = 10.5$$

$$SD_G = \sqrt{\frac{\sum x^2}{N}} = \sqrt{\frac{64.00}{8}} = \sqrt{8.00} \qquad SD_B = \sqrt{\frac{214.00}{8}} = \sqrt{26.75}$$

$$= 2.83 \qquad\qquad\qquad\qquad\qquad = 5.17$$

same, inspection of the raw scores suggests that the boys are more variable than the girls: the range of the boys' distribution, for example, is 15 while the girls' is 9. This greater variability is reflected in the SD: 5.17 score units for the boys vs. 2.83 for the girls.

Sample and Population SD

Before proceeding further with our discussion of SD, let us anticipate a topic that we will present in more detail in later chapters. In research we seldom can measure all the individuals or objects who have the characteristic with which we are concerned, and we therefore measure only a portion of the total group. If we wished to determine, for example, how well fifth-grade children in the United States could do on a geography test we have devised, we would find it impractical to try to administer the test to all current fifth-graders in the country. We therefore would measure only a *sample* of children from the total group or *population* of fifth-graders. Then we would proceed to describe the characteristics of the sample by computing such statistics as the mean and SD of the distribution. Rarely, however, are these sample characteristics of any interest to us in and of themselves. Rather, we wish to use the data obtained from samples to make guesses about the characteristics of the populations from which the samples were drawn.

In later chapters we will be paying a good deal of attention to the methods by which such estimates may be made. At present we will discuss the relationship between the sample and population values of the mean and of the standard deviation and variance. We should mention first that statisticians have found it convenient to use different symbols to refer to the characteristics of samples and populations. In subsequent discussions we will use a new symbol, the Greek letter μ (mu) to indicate the mean of a *population* and reserve the symbol $\overline{X}$, which we have used exclusively up to this point, to indicate the mean of a *sample* of scores from some larger population. The standard deviation of a *population* will be indicated by the small Greek letter σ (sigma). To indicate the SD of a *sample* we will place a tilde over the sigma: $\tilde{\sigma}$. The *variance* of a population and a sample thus will be identified as σ^2 and $\tilde{\sigma}^2$, respectively.

A sample $\overline{X}$ rarely has exactly the same value as the mean of its population (μ). However, assuming we have selected our sample in some unbiased manner (a topic to which we will address ourselves in later chapters), $\overline{X}$ shows no systematic tendency to be either larger or smaller than its μ. Thus, the mean of a sample is the best single guess we can make about the value of an unknown population mean. However, a sample SD ($\tilde{\sigma}$) tends to be *smaller* than the SD of its population (σ), so that if we were to use $\tilde{\sigma}$ as our guess about σ, we would be likely to underestimate the population value. Further, the degree to which sample SD's underestimate the population sigma tends to *increase* as sample size *decreases*. The same statements can also be made about the relationship between the *variance* of a sample ($\tilde{\sigma}^2$) and the population variance (σ^2).

In explaining why the sample SD and variance tend to be smaller than the population values, we will find it simpler to discuss the matter in terms of the variance. Let us consider first the components of the formula for the variance, namely $\sum x^2$ and N. As the N of a sample increases, not only the denominator of the formula but also the numerator, $\sum x^2$, increases. That is, as we add a case to N, we also have its x^2 to add to $\sum x^2$. Of course, if these two components increased proportionately, the variance would remain constant. This is pretty close to what happens but not quite. As we select more and more cases, it becomes increasingly likely that extremely deviant scores will be included within our sample. For example, we would be more likely to find a man over seven feet tall—or a man well under five feet—in a group of 500 men than in a group of twenty. These extreme cases add more to $\sum x^2$ than they do to N, simply because their deviations from the mean are bigger than those we have gotten for cases closer to the middle of the distribution. So we may expect some increase in the variance with increases in N. However, when N is relatively large to begin with, say thirty or forty cases, we expect less change in the variance with increases in N than when it is small.

Since a sample is, by definition, smaller than the population from which it was obtained, we can now see why a sample variance tends to be smaller than the variance of its population. To obtain an unbiased estimate of the population variance from sample data, we include a *correction factor* in the basic formula for the variance by dividing the $\sum x^2$ obtained from our sample by $N - 1$ rather than by N. It is customary to use the symbol s^2 to indicate an unbiased estimate of a population variance from the data of a sample. Thus:

$$s^2 = \frac{\sum x^2}{N - 1} \tag{5.2}$$

where, just as before, N is the number of cases in the sample and $\sum x^2$ is the sum of the squared deviations of the scores from the sample mean.

Since we lower the value of the denominator by subtracting one from the sample N, you can see that s^2, our estimate of the unknown population variance, will always be slightly larger than the variance of the sample ($\tilde{\sigma}^2$). Further, the difference between s^2 and $\tilde{\sigma}^2$ will be greater in samples with small N's than with large ones. This is as it should be, since the degree to which a sample variance underestimates the population variance is inversely related to sample size, so that a greater correction is required when small samples are used to estimate population variances than when large ones are used.

In order to obtain an estimate of the population SD from sample data, we simply find the square root of s^2. Thus, the formula for estimating the population sigma (σ) from a sample is:

$$s = \sqrt{\frac{\sum x^2}{N - 1}} \tag{5.3}$$

For a rather technical mathematical reason, s does not yield a completely unbiased estimate of the population value, even though s^2 does. The degree of bias that remains, however, is small.

By this time you may be confused by the various kinds of SD's (and SD^2's or variances) we have discussed and the different symbols we have used to refer to them. As you become experienced with problems involving samples and populations, the distinctions among them should become quite clear. We can at least try to hasten the process by stating once more the three kinds of SD measures we have presented and the formulas for each:

A. SD of a *population* of measures (σ):

$$\sigma = \sqrt{\frac{\sum x^2}{N}} = \sqrt{\frac{\sum (X - \mu)^2}{N}} \tag{5.4}$$

where μ is the population mean and N is the number of cases in the population

B. SD of a *sample* of measures from some population ($\tilde{\sigma}$):

$$\tilde{\sigma} = \sqrt{\frac{\sum x^2}{N}} = \sqrt{\frac{\sum (X - \overline{X})^2}{N}}$$

where $\overline{X}$ is the sample mean and N is the number of cases in the sample.

C. *Estimate* of an unknown population σ from the data of a sample (s):

$$s = \sqrt{\frac{\sum x^2}{N - 1}} = \sqrt{\frac{\sum (X - \overline{X})^2}{N - 1}} \tag{5.5}$$

where -1 in the denominator is the correction of the sample SD's underestimation of the population SD.

We return now to a consideration of SD as a purely descriptive index of the amount of variability among a group of individuals whom we have actually measured. For convenience, we will assume in our discussion that we are dealing only with *samples* from some population and hence use only the symbol $\overline{X}$ to refer to the mean of a distribution and $\tilde{\sigma}$ or SD to refer to its standard deviation.

FURTHER COMPUTATIONAL TECHNIQUES

SD from Frequency Distributions

In the illustration we have given of the computation of SD (Table 5.1) we used ungrouped data. We may also compute SD for data grouped into frequency distributions, just as we did in the case of $\overline{X}$ in Chapter 3. Such a problem is worked out in Table 5.2. The first three columns are those we use customarily to obtain $\overline{X}$. The fourth column, x, is the deviation of each raw score from $\overline{X}$, and the fifth column gives the squares of these deviation scores. In the final column each squared deviation is multiplied by the frequency with which it occurred, giving fx^2. The sum of this column gives the total of all fx^2 values. We find the variance by dividing this $\sum fx^2$ by the total number of cases and SD by taking the square root.

Table 5.2   Calculation of SD with grouped data and deviation
scores.

X	f	fX	x	x^2	fx^2
11	3	33	4.37	19.10	57.30
10	3	30	3.37	11.36	34.08
9	12	108	2.37	5.62	67.44
8	15	120	1.37	1.88	28.20
7	23	161	.37	.14	3.22
6	24	144	− .63	.40	9.60
5	13	65	−1.63	2.66	34.58
4	10	40	−2.63	6.92	69.20
3	5	15	−3.63	13.18	65.90
	N = 108	$\sum fX = 716$			$\sum fx^2 = 369.52$

$$\overline{X} = \frac{\sum fX}{N} = \frac{716}{108} = 6.63$$

$$\tilde{\sigma} = \sqrt{\frac{\sum fx^2}{N}} = \sqrt{\frac{369.52}{108}} = \sqrt{3.42} = 1.85$$

Raw-Score Method

The use of deviation scores (x) is fairly tedious and, because decimals usually result when $\overline{X}$ is subtracted from X to obtain x, inaccuracies due to rounding and calculational errors are likely to creep in. Instead of using the deviation method of computing SD's, it is usually more convenient to use what is commonly called a raw-score formula for calculating SD. This formula is as follows:

$$\tilde{\sigma} = \sqrt{\frac{\sum X^2}{N} - \left(\frac{\sum X}{N}\right)^2} \quad \text{or, since } \frac{\sum X}{N} = \overline{X}, \quad \tilde{\sigma} = \sqrt{\frac{\sum X^2}{N} - \overline{X}^2}$$

Origin of Raw-Score Formula.   It will be worthwhile to see just how this formula came about. Knowing this, you will have at least one formula that need not be taken on faith and you will get a little understanding of the derivation of formulas. We start, of course, with the basic formula for SD:

$$\tilde{\sigma} = \sqrt{\frac{\sum x^2}{N}}$$

The first step is to square both sides of the equation, obtaining the variance: $\tilde{\sigma}^2 = \sum x^2/N$. We know that $x = X - \overline{X}$, so we substitute this to get:

$$\tilde{\sigma}^2 = \frac{\sum (X - \overline{X})^2}{N}$$

Now, to expand the term, by squaring $X - \overline{X}$:

$$\tilde{\sigma}^2 = \frac{\sum (X^2 - 2\overline{X} X + \overline{X}^2)}{N}$$

Placing the summation sign and N with each term gives:

$$\tilde{\sigma}^2 = \frac{\sum X^2}{N} - \frac{2\overline{X} \sum X}{N} + \frac{\sum \overline{X}^2}{N}$$

Looking at the second term on the right-hand side, we see that a portion of it is $\sum X/N$, which we know equals $\overline{X}$. So, substituting $\overline{X}$ for $\sum X/N$, we get:

$$\tilde{\sigma}^2 = \frac{\sum X^2}{N} - 2(\overline{X})(\overline{X}) + \frac{\sum \overline{X}^2}{N}$$

which reduces to:

$$\tilde{\sigma}^2 = \frac{\sum X^2}{N} - 2\overline{X}^2 + \frac{\sum \overline{X}^2}{N}$$

Now, look at the numerator in the last term, $\sum \overline{X}^2/N$. We know that $\sum$ means summation from the first to the last score in a distribution, the sum of *N scores*. Thus $\sum \overline{X}^2$ says that we should sum $\overline{X}^2$ together N times, and the last term may be rewritten $N\overline{X}^2/N$. If these N's are cancelled, we have:

$$\tilde{\sigma}^2 = \frac{\sum X^2}{N} - 2\overline{X}^2 + \overline{X}^2$$

and then, $-2\overline{X}^2 + \overline{X}^2 = -\overline{X}^2$, so:

$$\tilde{\sigma}^2 = \frac{\sum X^2}{N} - \overline{X}^2$$

Taking the square root of both sides produces the formula for raw-score calculation of SD:

$$\tilde{\sigma} = \sqrt{\frac{\sum X^2}{N} - \overline{X}^2} \qquad \text{(5.6)}$$

The final step shall be to give concrete proof that this formula is equivalent to the basic formula for SD, $\sqrt{\sum x^2/N}$. In Table 5.3 we have used the raw-score method to compute the SD for the girls' data shown in Table 5.1. The SD (2.83) in Table 5.3 is exactly the same value we found by the deviation method in Table 5.1.

Table 5.3 Calculation of SD by the raw-score method.

X	X²
15	225
14	196
12	144
10	100
10	100
9	81
8	64
6	36
$\sum X = 84$	$\sum X^2 = 946$

$$\bar{X} = \frac{84}{8} = 10.5$$

$$\tilde{\sigma} = \sqrt{\frac{\sum X^2}{N} - \bar{X}^2} = \sqrt{\frac{946}{8} - (10.5)^2}$$

$$= \sqrt{118.25 - 110.25} = \sqrt{8} = 2.83$$

Grouped Data.   The raw-score formula may also be used with grouped data. All we do is add a frequency column to the work sheet. An illustration is worked out in detail in Table 5.4. These data show the varying number of pounds of butter wrapped in wax paper by a single operator during fifteen-minute work periods. The data are for 171 such work periods. In this table we have placed the *f* column before the X column so that our multiplication is always for adjacent columns. The main job is to get the $\sum fX^2$. To do this we first multiply each *f* by X (which in our example is the midpoint of the class), which will allow us to obtain the $\sum fX$ from which we get $\bar{X}$. Then we multiply each value in the X column by the corresponding value in the *f*X column. The result of multiplying (X)(*f*X) is *f*X². We sum these quantities to get $\sum fX^2$, which we substitute directly into the formula:

$$\tilde{\sigma} = \sqrt{\frac{\sum fX^2}{N} - \bar{X}^2} \tag{5.7}$$

Table 5.4 Calculation of SD by raw-score method with grouped data.

Class Interval	f	X	fX	fX²
62–64	1	63	63	3969
59–61	4	60	240	14400
56–58	23	57	1311	74727
53–55	47	54	2538	137052
50–52	47	51	2397	122247
47–49	30	48	1440	69120
44–46	14	45	630	28350
41–43	5	42	210	8820
	N = 171		$\sum fX = 8829$	$\sum fX^2 = 458685$

$$\overline{X} = \frac{\sum fX}{N} = \frac{8829}{171} = 51.63$$

$$\tilde{\sigma} = \sqrt{\frac{\sum fX^2}{N} - \overline{X}^2} = \sqrt{\frac{458685}{171} - (51.63)^2} = \sqrt{2682.37 - 2665.66}$$

$$= \sqrt{16.71} = 4.09$$

SOURCE: Data from Rothe, H. F. Output rates among butter wrappers: II. Frequency distributions and an hypothesis regarding the "restriction of output." *Journal of Applied Psychology*, 1946, 30, 320–327.

Raw-Score Formula for s.   It is also possible to use a raw-score formula to calculate *s*, a statistic which, you will recall, is an estimate of the population sigma based on sample data. This formula is as follows:

$$s = \sqrt{\frac{\sum X^2 - (\sum X)^2/N}{N - 1}} \tag{5.8}$$

The application of this formula may be illustrated by the data of Table 5.3, in which $\sum X$ and $\sum X^2$ have already been determined. Entering these values into the equation, we have:

$$s = \sqrt{\frac{946 - (84)^2/8}{8 - 1}} = \sqrt{\frac{946 - 882}{7}} = \sqrt{9.14} = 3.02$$

Since the formula for *s* attempts to correct for the fact that the $\tilde{\sigma}$ of a sample tends to underestimate the population sigma, the estimate of the population value we just obtained is, as expected, somewhat higher than the SD of the sample found in Table 5.3 (3.02 versus 2.83).

The raw-score formula for s can also be easily modified so that it may be applied to grouped data. $\sum fX$ and $\sum fX^2$ are found in the usual way and substituted in the formula for $\sum X$ and $\sum X^2$.

z SCORES

We now know what the SD is: the square root of the mean of the x^2's. The SD is thus a rather special kind of "average," telling us the "average" number of *score units* by which individuals deviate from the mean. Once we have calculated the SD of the distribution, is there any way we can picture or visualize it? Can we find any additional use for SD besides describing a distribution's variability? Again, we will work up to the answers slowly.

The distribution of final exam scores in a large biology class turns out to have a mean of 65 and an SD of 10. Thus, the "average" amount by which individuals deviate from $\overline{X}$ is 10 units. Lou has a score of 85. Is that good or bad? We see that Lou's score is 20 points above the mean ($x = 85 - 65 = 20$); Lou deviates from the mean twice as much as the "average" individual. With a score of 85, Lou did quite well indeed. How about Pat with a score of 67? Pat's performance was middling, slightly above the mean but deviating from it by only 2 points ($x = 67 - 65 = 2$). This deviation is far less than an "average" amount—to be precise, only $^2/_{10}$ of an average amount. Finally, what about Bill, with a score of 35? We can only conclude that Bill bombed biology. With a score 30 points below the mean, three times worse than the "average" deviation, Bill's performance was obviously abysmal.

What we have done in each of these examples is to use the SD as a kind of measuring unit, translating each raw score into an "SD score." Lou, with a raw score of 85, had an SD score exactly 2 units above the mean (a deviation of 20 points from the mean is twice the SD of 10). Bill, with a raw score of 35 and x of 30 ($x = 65 - 35$), scored 3 SD units below the mean. Pat's score of 67 fell $^2/_{10}$ of an SD unit above the mean of 65. We could, of course, take *any* raw score from the distribution and express its position in SD units. What we have created, then, is an alternative to the raw-score scale, using the SD as the unit of measurement instead of the raw score.

This is not our first discussion of an alternative to the raw-score scale. In previous chapters we discussed using percentiles for this purpose and methods for translating raw scores (points on the raw-score scale) to percentiles (points on the percentile scale) and back again. We can also formally describe a method that permits us to go back and forth between the raw-score scale and the SD scale. Before doing so, let us see how the SD scale can be pictured and how to lay one out.

Figure 5.2 shows a frequency polygon whose mean has been calculated to be 25 and whose SD is 5. As usual, the baseline of the polygon shows raw

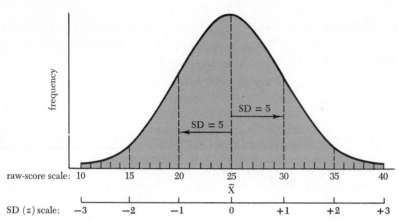

Figure 5.2   Raw-score scale and SD or z-score scale for a distribution in which X̄ is 25 and SD is 5.

scores; that is, the baseline represents a *scale*, divided into steps that have a width of one raw-score unit. Below the raw-score scale we have placed the SD scale, whose steps have a width of *one SD unit*. Exactly how wide is this SD unit? It is, of course, as wide as SD has been calculated to be for a given distribution, and it is expressed in raw-score units. In Figure 5.2, in which SD is 5, one SD unit is therefore 5 raw-score units wide. (This is comparable to saying that a foot is 12 inches wide.)

The SD scale is also known as the *z* scale and the scores on it as *z* scores. In marking off the SD or *z* scale, we start at the *mean* of the distribution and work in both directions. In the distribution in Figure 5.2, which has a mean of 25, we therefore start at the point that is equivalent to 25 on the raw-score scale. The SD is 5 units. In laying off the upper portion of the scale, we therefore go up from X̄ to a point that is equivalent to 5 units on the raw-score scale. The distance between the mean and this point is one SD unit, and the point has a value, in SD or *z* units, of +1. Going up the scale the equivalent of 5 more raw-score units brings us to an SD score of +2, and another 5 units to +3. The raw-score equivalents of these three *z* scores, we can see, are 30 (X̄ + SD = 25 + 5), 35 (X̄ + 2SD = 25 + 10), and 40 (X̄ + 3SD = 25 + 15).

Notice in Figure 5.2 that almost all of the cases that lie above X̄ fall between the X̄ and an SD or *z* score of +3. Assuming that the total N is large, this is typically true of all symmetrical bell-shaped distributions, such as the idealized curve shown in the figure. Thus, at least when N is some finite number, approximately three SD's will cover all or almost all the distance between X̄ and the highest score in the distribution.

We use the same procedure to mark off a succession of SD widths below $\overline{X}$. Going 5 raw-score units below $\overline{X}$ brings us to an SD or z score of -1, 10 units to a z score of -2, and 15 units to -3. Since the distribution is symmetrical, we are not surprised to find that three SD's also cover the distance between $\overline{X}$ and the lowest raw score in this sample of measures.

We are now aware that SD is a unit of width that is some multiple of the raw-score width. For any given distribution SD is a calculable value of fixed size, but it may vary from one distribution to another. If R in one distribution is 24 and in another 48, we would not expect SD to be the same size for both. Now, we are accustomed to think of our measuring scales for linear dimensions such as width to be fixed and unalterable. A yard is a yard and a meter is a meter, and neither varies in size. In the case of SD as a width measure, however, we must change this conception. To repeat, SD is a given value for a specific distribution, but may vary from distribution to distribution.

Finding z from X

We are now ready to discuss the procedures we use to translate back and forth between the scales, finding the z-score equivalent of raw scores and vice versa. Take first a simple and familiar example: if we want to translate feet into yards, we divide the number of feet by three. If we want to change ounces to pounds, we divide the number of ounces by 16. In the same fashion, if we want to change a raw score into a z score, we find the distance that score is from $\overline{X}$ and divide it by SD. In Figure 5.2, for example, where $\overline{X} = 25$ and SD $= 5$, what is the z-score equivalent of a raw score of 28? This score lies 3 raw-score units above $\overline{X}$ ($28 - 25 = +3$); dividing this distance by the SD of 5 gives us a z score of $+ 60 \longrightarrow \dfrac{3}{5}$. If we put these steps into equation form, we have:

$$z = \frac{X - \overline{X}}{\tilde{\sigma}} \qquad\qquad (5.9)$$

But, since $X - \overline{X}$ is the definition of x, we can also write the formula as:

$$z = \frac{x}{\tilde{\sigma}}$$

Let's work out a few examples from a distribution with a mean of 42 and SD of 8. For a raw score of 54:

$$z = \frac{X - \overline{X}}{\tilde{\sigma}} = \frac{54 - 42}{8} = \frac{12}{8} = +1.5$$

If the raw score is below $\overline{X}$, the z score takes a minus value. For example, in the same distribution, what is the z score of a raw score of 36?

$$z = \frac{36 - 42}{8} = \frac{-6}{8} = -.75$$

Finding X from z

We have just seen how to translate a raw score into a z score, expressed in SD units. We can reverse this procedure, so that if we know a z score we can obtain a raw score. We could use exactly the same formula as before and solve for an unknown raw score. However, it is more convenient to rearrange the formula so that the raw score X becomes the unknown on the left-hand side of the equation. This formula is:

$$X = \overline{X} + z(\tilde{\sigma}) \tag{5.10}$$

For example, with $\overline{X} = 42$ and $\tilde{\sigma} = 8$, what is the raw-score equivalent of a z score of -1.5? Applying Formula (5.10), we find:

$$X = 42 + (-1.5)(8) = 42 - 12$$
$$= 30$$

We do not change raw-score distances into SD distances and back again merely to get practice in arithmetic. These z scores have very definite uses. For example, the need sometimes arises to compare the same person with himself in two different distributions of scores. If a group of men take a test of clerical aptitude and a test of mechanical aptitude, we might want to know the relative standing of an individual on both tests. If both tests have the same $\overline{X}$ and SD, and the distributions are similar in shape, the raw scores can be compared directly. This would be a rare case. More likely, the two distributions will have different $\overline{X}$'s and SD's. Thus, a particular raw score on one test will probably mean something quite different from the same raw score on another.

One satisfactory way of comparing an individual across distributions is by comparing ranks. This, in effect, is what a percentile rank is. If a person has a percentile rank of 78 on one test and of 62 on the other, we know he has done better on the first test than on the second. We may use z scores for this same purpose of comparing two scores in distributions in which the $\overline{X}$'s and SD's are different. Suppose that a group of men are given the mechanical and clerical aptitude tests as suggested above, with the following result:

	$\overline{X}$	SD
Mechanical	100	10
Clerical	60	6

A man gets a raw score of 69 on the clerical test and 75 on the mechanical test. His z score for the former is:

$$z = \frac{x}{\tilde{\sigma}} = \frac{9}{6} = +1.5$$

and for the latter:

$$z = \frac{-25}{10} = -2.5$$

He is much worse on the mechanical aptitude test than on the clerical test, even though his raw score is higher for the mechanical test. That is, his z score on the mechanical test is nearly at the bottom of the distribution, 2.5 SD's below $\overline{X}$, while it is considerably above $\overline{X}$ on the clerical test. The pattern of aptitude shown by this individual suggests that he might do well in a job requiring clerical skills but ought to leave anything mechanical to other people such as his wife.

Table 5.5 shows two distributions of test grades obtained from the same eighteen graduate students in psychology. One test was an examination in advanced statistics and the other was concerned with principles of psychology. The range and SD are about the same for the two tests, but the $\overline{X}$'s are quite different. Comparing raw scores does not mean very much. For example, the first student earned about the same raw score on both tests. However, on the principles test he is about one-half SD below $\overline{X}$ and on the statistics test he is above $\overline{X}$. Student number six earned quite different raw scores but about the same z scores. In brief, by translating raw scores to z scores, we can easily make direct comparisons of position in the group.

A CHECK ON CALCULATIONAL ERRORS: SD AND R

In discussing z scores, we noted that the sizes of R and SD are related. This suggests that we can check on computational errors by comparing the calculated SD with the range (R) of scores in the distribution. Students may make enormous arithmetic errors in calculating SD, and it is worthwhile exploring the relationship between R and SD as a means of avoiding them. We have already seen that in a sample with a large N and a symmetrical bell-shaped distribution, such as the one in Figure 5.2, all or most of the scores are encompassed by six SD units, three on either side of the mean. For large samples that approximate this curve shape, therefore, you can check your calculated SD to see if it is approximately one-sixth of R. After you have calculated several SD's you will begin to get the "feel" of them in relationship

Table 5.5 Calculation of z-scores on two tests given to 18 graduate students.*

Student Number	X	Principles X − X̄	z	X	Statistics X − X̄	z
1	66	− 6.17	− .55	65	3.67	.29
2	79	6.83	.61	71	9.67	.76
3	86	13.83	1.23	72	10.67	.84
4	58	− 14.17	− 1.26	44	− 17.33	− 1.36
5	75	2.83	.25	63	1.67	.13
6	69	− 3.17	− .28	57	− 4.33	− .34
7	100	27.83	2.47	82	20.67	1.62
8	66	− 6.17	− .55	71	9.67	.76
9	64	− 8.17	− .72	45	− 16.33	− 1.28
10	65	− 7.17	− .64	46	− 15.33	− 1.20
11	73	.83	.07	61	− .33	− .03
12	54	− 18.17	− 1.61	73	11.67	.92
13	62	− 10.17	− .90	35	− 26.33	− 2.07
14	80	7.83	.69	64	2.67	.21
15	81	8.83	.78	62	.67	.05
16	60	− 12.17	− 1.08	50	− 11.33	− .89
17	80	7.83	.69	62	.67	.05
18	81	8.83	.78	81	19.67	1.55

$$\bar{X} = 72.17 \qquad\qquad \bar{X} = 61.33$$
$$R = 46 \qquad\qquad R = 47$$
$$SD = 11.27 \qquad\qquad SD = 12.73$$

* One test measured knowledge of principles of psychology; the other advanced statistical knowledge.

to the distribution of scores from which they have been calculated. You will quickly find out that if R of your distribution is 24 and your calculated SD is 25, something is wrong with your calculations, since SD is larger than R. Or, if you have R of 24 and you obtain SD of 1, something also is wrong. But we must stress now, and stress emphatically, that we do not calculate SD by taking one-sixth of R; rather, *after* we have calculated SD we check to see if its value is approximately one-sixth of R. This provides only a very crude check, unless the number of measures is very large and the distribution quite symmetrical and shaped like Figure 5.2. However, so many sets of data are distributed like those of Figure 5.2 that we will find our check has considerable usefulness.

We have said that the ratio of 6:1 between SD and R holds only when N is large. When N is small, there will be less than six SD's in R. To understand this, recall that the greater the number of cases we draw, the more

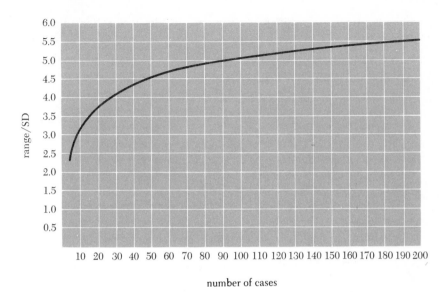

Figure 5.3   Curve for making a rough check on accuracy of calculation of SD.

likely we are to get extreme scores (those at the tails of the distribution). The addition of extreme scores increases R quite markedly. The SD also increases, as we discussed earlier, but to a lesser degree. Now, if R tends to increase fairly sharply as N increases and SD increases less sharply, the ratio between R and SD must increase as N increases. That is, R/SD increases with increases in N. With large N's this R/SD ratio is about six and is a useful rough check on our computational accuracy. We may also have such checks when N is not large. Figure 5.3 gives the number of SD's in R (R/SD) to be expected for N's from 5 to 200.[1]

This is the way to use Figure 5.3. Suppose we calculate SD for a fairly symmetrical distribution of 30 cases and get SD of 5. We note that R is 20. Hence, R/SD is 4. Now, we look on the abscissa (horizontal axis) at 30, look directly above until we hit the curve, then cross over to the ordinate where we see that 4 is about the proper ratio for this N. Take another illustration. With 90 cases, and R of 40, we get SD of 3. Something is wrong; the ratio should be approximately 5, whereas from our calculation it is about 13.

We must add a final word of caution about estimating SD from R. The check on correctness of SD provided by Figure 5.3 is crude; if the calculated SD is 14 and Figure 5.3 indicates that it should be 15 or 16, we

[1] Tippet, L. H. C. On the extreme individuals and the range of samples taken from a normal population. *Biometrika*, 1925, 17, 364–387.

should not get upset. The difference between our calculated SD and estimated SD would have to be greater before we would suspect an error in arithmetic. There is, of course, no way to tell exactly how much difference there must be between calculated and estimated SD before we should suspect a computational error. But the use of Figure 5.3 will help us avoid gross errors, and that is all it is intended to do.

Up to this point we have discussed a series of *descriptive* statistics— techniques that permit us to describe the properties of groups we have observed. We now turn to *inferential* statistics—techniques that permit us to make estimates about population characteristics based on the data from the samples we have observed. Fundamental to inferential statistics is the notion of *probability*, which we explore in the next chapter.

TERMS AND SYMBOLS TO REVIEW

Range	Population mean (mu or μ)
Deviation score (x)	Population SD (σ)
Average deviation	Sample SD ($\tilde{\sigma}$)
Standard deviation (SD)	Estimate of population SD (s)
Variance (SD^2)	Raw-score method
Sample	z score
Population	SD unit

Probability

6

PROBABILITY AND STATISTICAL INFERENCE

The characteristics of the particular group or groups of individuals measured in scientific investigations are seldom of interest in and of themselves. Typically, our purpose is to discover something about a *population*, and we have included in our investigation only a sample of the individuals or events belonging to that population. In obtaining the mean of a sample, for example, we may hope to gain information about the mean of the population as a whole.

Even if samples are drawn from the population in an unbiased manner, the statistics obtained from a number of independent samples (for example, the mean of each) seldom agree exactly. Since the samples differ, no single sample can automatically be assumed to reflect exactly the properties of the population. Most samples are reasonably similar in their properties to the population, but on occasion they may be highly deviant.

Consider the following example. If a perfectly balanced coin is tossed a very large number of times (technically, an infinite number of times), a head (H) will come up half the time and a tail (T) the other half. (That H and T occur equally often is in fact our *definition* of a perfectly balanced or *fair* coin.) Now, suppose that instead of obtaining the entire *population* of tosses of a fair coin (possible only in theory anyway, since the population is infinitely large), we obtain a *sample* of only 10 tosses, tallying the total number of H and T for the series of 10. We repeat this procedure over and over again, gathering data on a substantial number of samples of 10 each. What do we expect the number of H and T to be in these samples? Our best *single* guess is 5 H and 5 T. But we wouldn't be at all surprised to find unequal splits, such as 6 H and 4 T or 3 H and 7 T, occurring quite frequently. Very occasionally, samples as deviant as all H or all T may occur. Samples, then, do not always mirror faithfully the population from which they were drawn.

Statistical theory does permit us, however, given certain facts about the population, to state the exact frequency with which events can be expected to occur. These same theoretical principles also permit us to describe the likelihood or *probability* of occurrence of some event or series of events. For example, if you toss an unbiased, balanced coin a single time, what is the probability of occurrence of H? Conversely, what is the probability of T? You know that these two events are *equally likely*, so you might describe H and T as having a 50-50 chance of occurring. Setting the sum of the probabilities of all possible events (in this case, of H and T) at 1.00, we therefore can describe the probability of H as .50 and of T as .50. Suppose that you are about to toss the coin 5 times. What is the probability of getting 3 H and 2 T? You will discover in later discussion that, over the long run, 3 H and 2 T can be expected to occur 10 times out of 32 or approximately 31 percent of the time. The theory of probability allows us to state that the probability of getting 3 H and 2 T in a *single* series of 5 tosses therefore is .31.

Now consider another example. We have a coin of unknown origin and ask whether it is a balanced, unbiased coin or a biased one. That is, we want to know whether an infinite number of tosses of the coin (the entire population of tosses) would produce half H and half T, or some other ratio. Not having infinite patience or time, we obtain instead a sample of only 50 tosses, finding 22 H and 28 T. Is the coin unbiased? No conclusive answer can be given, but we can make a reasonable guess by determining the probability that a sample with this H-T split or an even more extreme split would occur *if the coin were in fact unbiased.* If the probability is low—that is, the likelihood is very small that a sample this different from an even split would occur if the coin were unbiased—then we reject our assumption about the coin's fairness and conclude instead that it is biased. Conversely, if the probability is high of getting a result this extreme or more so with a fair coin, we conclude that this particular coin is likely to be unbiased. What we have done in this example is to use probability theory to *test a hypothesis* about a population character-istic and to reach some conclusion about it. Probability theory thus permits us to make reasonable guesses or inferences about populations, based on the data from samples.

In this chapter we will review some of the elementary principles of probability to lay the foundation for discussion in later chapters of hypothesis testing and the statistics of inference. It is a time-honored tradition in discussing probability to use simple, understandable examples such as coins, dice, cards, or balls in a bowl, and we have already begun our discussion in this way. Although these illustrations may be of interest only to gamblers, they are useful explanatory devices. With occasional exceptions we will therefore continue to employ them throughout the chapter.

SETS AND PROBABILITIES

We start with an example: we take five balls and number them 1, 2, 3, 4, and 5. These numbered balls constitute what is called a *set* and the individual balls the *elements* of the set. In this example the elements of the set are *mutually exclusive.* That is, each ball has a single number on it, so that if we put the balls in a bowl, mix them thoroughly, and then draw one out, so that only one number can occur. If the ball we select has a 3 on it, for example, it cannot simultaneously have any other number. (Some events are *not* mutually exclusive. For example, suppose we asked a group of individuals whether they used public transportation or some other means of getting to work yesterday. Some might say they had used a combination of both, such as driving their car to the railroad station and taking a train from there. Use of one form of transportation on the way to work did not preclude use of other forms. Similarly, each of our five balls could have more than one number on it from the set of numbers from 1 to 5. If we wished to classify each ball as having the property 1, 2, 3, 4, or 5, a ball having the numbers 2 and 3, for example, could be simultaneously classified as having the property 2 and 3; the set of properties in this instance is not mutually exclusive.)

Returning to our mutually exclusive set of five numbered balls (i.e., the balls with single numbers on them), suppose we went many, many times through the procedures of drawing a ball, replacing it, mixing up the balls, and drawing again. How frequently would we expect each ball to be selected? Before answering, we should note an additional characteristic of this procedure that is critical: *on any occasion, each element is equally likely to occur.* That is, except for the numbers, the balls are identical; and if we scramble them thoroughly no ball is any more likely than another to come to hand when we draw. We can therefore confidently state that over the long run, each ball will be selected $^1/_5$ of the time. We can also say that on the average each ball will be selected 20 percent of the time or that the average *proportion* of the time it will be selected is .20.

Now if we draw a single ball, how likely (how probable) is it that a given number will occur? What is the probability that the ball is, say, a 4? The answer follows from our conclusion that each ball will be selected $^1/_5$ of the time. That is, over repeated occasions, we expect a 4 to occur $^1/_5$ of the time. For a single draw, therefore, there is one chance in five of getting a 4; alternately, we can say that a 4 will occur 20 percent of the time on the average, or that it has a *probability* of occurrence of .20. What is the probability that in a single draw from our set of five balls we will pick a ball with an *even* number (i.e., a ball from the *subset* of balls numbered 2 and 4)? We have 2 chances out of 5 of drawing an even number, so the answer is $^2/_5$ or .40. What is the probability of *not* drawing an even number? We have divided the set into two nonoverlapping subsets, two balls with even numbers and three with odd numbers. The probability of drawing noneven (that is, odd) numbered balls is therefore $^3/_5$ or .60. What about the probability of a ball with either an odd number *or* an even number? Since the question encompasses all elements of the set, the probability has to be unity: $^5/_5$ or 1.00.

We offer one more example. Suppose we have ten balls, five of them numbered 1, three of them 2, and two of them 3. Each of the individual *balls* (that is, each of the elements of the set) has an equal probability of being selected on a single draw ($^1/_{10}$), but what about each *number*? Since there are five balls numbered 1, $^5/_{10}$ or half of them fall into this "1" category; the probability of drawing a 1 therefore is .50. Similarly, the probability of a 2 is $^3/_{10}$ or .30, and of a 3, $^2/_{10}$ or .20.

Probability Defined

With these examples before us we can now offer a definition of probability *for equally likely events*. When a single, random observation from a set is to be made, the *probability of an event from a specific subset of events is the ratio of the number of events belonging to that subset to the total number of possible events*. This definition can be expressed in a simple formula, in which n_A stands for the number of events belonging to subset A and N for the total number of all possible events:

$$p(A) = \frac{n_A}{N} \tag{6.1}$$

Thus, in our example of five numbered balls, three events go to make up the subset of balls with odd numbers (1, 3, 5); the probability of occurrence of one of these three events on a single draw (the probability of drawing an odd number) therefore is $^3/_5$ or .60. One event goes to make up the subset of balls with the number 4 in our example; the probability of occurrence of an event from this "4" subset on a single draw therefore is $^1/_5$ or .20, and so on.

We can also express N as $n_A + n_{\bar{A}}$, where $\bar{A}$ stands for all categories of the elements in the set that are *not* in the A category. ($\bar{A}$ is technically called the *complement* of A.) Thus, we can also write:

$$p(A) = \frac{n_A}{n_A + n_{\bar{A}}}$$

The probability that A will not occur—that is, $p(\bar{A})$—is $n_{\bar{A}}/(n_A + n_{\bar{A}})$, and the two probabilities, $p(A)$ and $p(\bar{A})$, must add up to 1.00.

Notice that the definition above is of the probability of *equally likely events*. The condition of equal likelihood does not always hold. Take the simple case of a "loaded" coin—a coin that does not have perfect balance so that H and T do *not* come up equally often. For example, a coin may be biased in a manner that results in 60 percent H and 40 percent T over an infinite number of tosses. The likelihood of H on a given occasion thus is greater than the likelihood of T, specifically .60 vs. .40. We may give a more general definition of probability that holds for this as well as for the equal-likelihood case. *The probability of a random event of a certain type occurring on a given occasion is the proportion of times it would occur on an unlimited number of occasions.*

Both the general and more restricted definitions of probability assume that we know all the possible outcomes (N) in the population of events as well as the proportion of occasions on which the specified category of events will occur. In real-life situations we frequently do not have sufficient information about the population to fulfill these assumptions. For example, what is the probability that a woman, asked to name her favorite color, will pick red? Not having questioned all women, we have no idea of the number of colors that might be named as preferences nor the frequency of each preference. But we could choose a *sample* of women in an unbiased manner and use their responses as *estimates* of the values in the population as a whole. In our subsequent discussion of the principles of probability, however, we will use illustrations based on theoretical populations, such as games of chance, or empirical populations (that is, actual groups that are limited in number) that we have been able to observe in their entirety. Our discussion will also be limited to examples of equally likely events.

SOME PRINCIPLES OF PROBABILITY

The Addition Rule

We have a deck of 52 well-shuffled cards and pick one at random. The probability of selecting an ace of hearts is $1/52$; similarly, the probability of selecting a king of spades is $1/52$. What is the probability that the card will be either an ace of hearts *or* a king of spades? The answer is the sum of the probabilities of the two individual events, $1/52 + 1/52$, or $1/26$. Similarly, what is the probability of an ace or a king of any suit? Since there are four aces and four kings in the deck, the answer is $4/52 + 4/52$ or $2/13$. These examples are illustrations of the *addition rule: in a set of mutually exclusive random events the probability of occurrence of either one event or another event (or subset of events) is the sum of their individual probabilities.* For the two-event case, this rule may be expressed.

$$p(\text{A or B}) = p(\text{A}) + p(\text{B}) \tag{6.2}$$

The formula may be expanded to include any number of events. For example, the probability of one of three events (or subsets of events), A or B or C, would add $p(\text{C})$ to the right-hand side of the equation. The probability that a card drawn at random would be an ace, king, or jack therefore is $^4/_{52} + {}^4/_{52} + {}^4/_{52} = {}^3/_{13}$.

The Multiplication Rule

In the previous section we were concerned with specifying the probability that on a single occasion one of two or more mutually exclusive events or subsets of events will occur (for example, an ace *or* a king). We now consider the probability of occurrence of a *series* of events. Suppose we toss a fair coin two times; what is the probability that on the first toss a head will turn up and on the second a tail? Observe first that in two tosses four patterns are possible: HH, HT, TH, TT; over the long run, two tosses of the coin will produce our specified pair of events, HT, $^1/_4$ of the time. The probability of obtaining HT on a single pair of tosses therefore is $^1/_4$ or .25. Similarly, what is the probability of obtaining HTH, in that order, in three tosses? With three tosses eight different patterns are possible: HHH, HTH, HHT, HTT, THH, TTH, THT, TTT. The probability of HTH is therefore $^1/_8$. For four tosses there are 16 possibilities, so that the probability of any one of them is $^1/_{16}$, and so on.

Application of the *multiplication rule* allows us to determine these probabilities without having to figure out and then count all of the possibilities. This rule states that *the probability of two or more independent events occurring on separate occasions is the product of their individual probabilities.* For the case of two independent events this rule can be expressed as:

$$p(A, B) = p(A) \times p(B) \tag{6.3}$$

where $p(A, B)$ is to be read as the probability of both A and B occurring.

Let's see how the rule works. In the case of a coin, H and T each have a probability of $^1/_2$ on a single toss. The pattern HT about which we inquired should thus occur $^1/_2 \times {}^1/_2 = {}^1/_4$ of the time. For three tosses the probability of a given pattern is $^1/_2 \times {}^1/_2 \times {}^1/_2 = {}^1/_8$, and so on. How about this example: what is the probability of drawing from a deck of 52 cards a 2-spot, followed (after replacing our first card in the deck) by a heart? Since a 2 has a probability of $^4/_{52}$ or $^1/_{13}$ and a heart a probability of $^{13}/_{52}$ or $^1/_4$, the answer is $^1/_{13} \times {}^1/_4 = {}^1/_{52}$.

Note that our definition of the multiplication rule states that the events whose probability of occurrence we are specifying must be *independent*. By independent we mean that the occurrence of one event is not conditional upon (does not change the probability of) the occurrence of any other event. Tossing a fair coin, for example, and observing a head does not mean that on the next toss a second head is any more or any less likely to occur; its probability remains $^1/_2$.[1] This independence can be demonstrated by tossing a coin many times and comparing the proportion of heads following heads and following tails. The proportion of H following both H and T, you will discover, remains $^1/_2$.

Conditional Probability

Independent Events. Suppose we have one hundred balls, numbered 1 through 100. We draw a ball numbered 3 and replace it. We now inquire about the probability that on the next draw we will select the ball numbered 68. This type of question concerns a *conditional probability*: the likelihood that an event will occur, given the fact that another event or series of events *has already occurred*. A conditional probability is indicated by the expression $p(B \mid A)$, which is read as "the probability of B given A." In our example, 3 is event A and 68 is event B.

The answer to our question above is that $p(B \mid A) = {}^1/_{100}$. With replacement of each ball before drawing another to keep the N at 100, the probability of selecting a specific ball remains $^1/_{100}$, no matter what ball or how many balls have previously been selected. That is, as we discussed in the section above, the events are independent; when we replace each ball before drawing another, the probability of the individual event remains constant whatever prior events have occurred. When events are independent, you notice, the probability of B given A $[p(B \mid A)]$ reduces to $p(B)$. When events are *not* independent, $p(B \mid A)$ is *not* equal to $p(B)$, as we shall see in the next section.

Nonindependent Events. We now turn to the conditional probabilities of nonindependent events. Suppose, to relieve the tedium of our previous examples, we imagine that six men on a hunting trip find themselves trapped in a mountain cabin by a snowstorm. Three of them happen to be lawyers, two are physicians, and one is an architect. One of them, they decide, should

---

[1] We are here denying what is known as the Gambler's Fallacy. Many gamblers believe, for example, that if a head has occurred with unusual frequency, a tail is "overdue" and therefore has a higher probability of occurrence than .50. In actual fact, no matter how many heads happen to come in a row—5, 10, or 200— the probability of tail on the next toss of a fair coin stubbornly remains at .5.

The Gambler's Fallacy is illustrated by Doug Krikorian's comments in the Los Angeles *Herald Examiner* (December 6, 1974, p. C-1): "It was pointed out . . . that the Rams would have a better chance to beat [the] Washington [Redskins] in their probable December 22 playoff opener if they lost to the Redskins Monday night. The feeling being, of course, that the law of averages would favor a Ram split rather than a sweep over the Redskins in a 13-day span." A football team's successive games, of course, are not completely independent. But the implication that because a split in two games was more likely for these two teams than a sweep, a Ram win in the first game would make a Ram loss in the second game more likely, is an obvious misunderstanding of the "law of averages."

go for help, and they decide to draw lots for the honor. The man chosen is one of the lawyers (the most likely event, since the probability of selection of the architect is $^1/_6$, of one of the physicians $^2/_6$, and of one of the lawyers $^3/_6$). After days of waiting, the five remaining men decide to send out a second messenger. What is the probability that the person selected by their second lottery will be a lawyer? With one of the lawyers gone and the group diminished to 5, the probability of another lawyer's being selected as the second man is $^2/_5$. If, instead, the *architect* had gone first, the probability of a lawyer's going second would be $^3/_5$. Or suppose that we were concerned with the probability that a *physician* went second, given that a physician had also gone first. In this instance the probability would be $^1/_5$.

These are examples of *conditional probabilities for nonindependent events.* Questions are being asked about $p(B \mid A)$, the likelihood of B given A, in a situation in which event A is not replaced in the original set before selection of B (the first man didn't come back). Without replacement, the probability of B *is* influenced by what has gone before. The reasoning by which $p(B \mid A)$ is determined is no different from that by which $p(A)$ is determined, except that we must take into account the change in the set brought about by prior events that have not been replaced. For example, if the balls numbered 2 and 4 are removed from a set numbered 1 through 100 and are not replaced, what is the probability that any particular ball of the remaining 98 will be selected on the next draw? The answer is $^1/_{98}$. Since the balls that were removed were both even, the probability of an *even* number on the next draw is $^{48}/_{98}$, and so on.

In the example of the hunters above, what is the probability that the first and the second men will both be lawyers? In a set of balls numbered 1 through 10, what is the probability, assuming nonreplacement, that the first ball will be 2, the second 4, and the third 5? These questions about a series of events are subject to the same sort of *multiplication rule* governing independent events. This rule, we recall, states that the joint probability of a series of events is the product of the probabilities of the individual events. Our previous statement of this rule, however, has to be modified to take into account that the probability of event B is conditional upon event A, C is conditional upon B, and so on. Thus, for three events (A, B, C) we have: $p(A, B) = p(A) \times p(B \mid A)$, so that $p(A, B, C) = p(A, B) \times p(C \mid A, B)$ and:

$$p(A, B, C) = p(A) \times p(B \mid A) \times p(C \mid A, B) \qquad (6.4)$$

where $p(A)$ is the probability of A, $p(B \mid A)$ is the probability of B given A, and $p(C \mid A, B)$ is the probability of C given A and B in that order.

The answer to our question about the two hunters being lawyers is therefore $(^3/_6)(^2/_5) = ^1/_5$ (or .20). The probability of drawing balls 2, 4, and

5 in that order without replacement from the set of 10 is $(^1/_{10})(^1/_9)(^1/_8)$ = $^1/_{720}$ (or .0014), and so on.

Conditional Probabilities and Joint Events

We now extend our discussion to consider conditional probabilities in situations in which each member of a group has been categorized according to two characteristics rather than one. Suppose, for example, that you are trying out as a ship's recreation director on a weekend cruise for "singles." You have been told that if the cruise is a success, you can have the job permanently. Quickly you size up the 80 men and 120 women who are on the cruise and assign each to one of three groups: Swingers, who are going to have a good time anyway, Duds, for whom little can be done, and Eagers, who are ready to enjoy themselves, given a little encouragement. You decide to concentrate your efforts on the Eagers. (For convenience, we will assume the validity of your judgments. We will also assume that for your purposes—getting the job—the group of 200 constitutes a complete *population* of individuals.) The results of your assignments are shown in Table 6.1.

Table 6.1 Number of men and women classified in the three Enjoyment-Potential groups.

	Swingers (A_1)	Eagers (A_2)	Duds (A_3)	Total
Men (B_1)	25	30	25	80
Women (B_2)	25	60	35	120
Total	50	90	60	200

Members of the group, you will notice, have been classified according to *two* characteristics: first, their Enjoyment Potential, in which each of the 200 individuals is assigned to one of three mutually exclusive categories (A_1, A_2, A_3), and second, their sex, in which each is assigned to one of two mutually exclusive categories (B_1, B_2). The entries in the cells of the table indicate the number of individuals exhibiting each pair of characteristics. The B_1A_3 cell, for example, shows that 25 of the 80 men are classified as Duds. These numbers can also be expressed as the *proportion* of the total group exhibiting each pair of characteristics. The frequency data for the group of 200 men and women just reported, when translated into proportions, yield the values shown in Table 6.2.

Table 6.2 Proportion of the 200 individuals falling into each category.

	Swingers (A_1)	Eagers (A_2)	Duds (A_3)	Total
Men (B_1)	.125	.150	.125	.400
Women (B_2)	.125	.300	.175	.600
Total	.250	.450	.300	1.000

Table 6.2 is also a *probability table*. For example, the probability that an individual drawn at random will be a woman (fall in the B_2 category) is .60, the sum of the three A entries for B_2. Similarly, the probability that an individual chosen at random is a Swinger (A_1) is .25, the sum of the two B entries for A_1. We could also inquire about the probability that an individual will jointly exhibit a given pair of characteristics. For example, what is the probability that a randomly selected individual will be both female *and* a Swinger (will be classified as A_1B_2)? We can read this probability directly from the table by locating the entry in the appropriate cell; we find it is .125.

Now we ask a somewhat different question. Given that an individual selected from the population of 200 is a *woman* (belongs to category B_2), what is the probability that she is an Eager (belongs to category A_2)? You will recognize this question as inquiring about a *conditional* probability, in this instance $p(A_2 \mid B_2)$ or the probability of A_2 given B_2. In attempting to determine this conditional probability, let us first go back to the frequencies in Table 6.1. There are 120 women in the group, 60 (50 percent) of whom are classified as Eager. Given that an individual is a woman, the probability that she falls into the Eager category is therefore .50. Among the males, the probability that a man is a Dud is $^{25}/_{80}$, and so on.

This type of conditional probability can also be determined by the following formula, using the probability values shown in Table 6.2:

$$p(A \mid B) = \frac{p(A, B)}{p(B)} \tag{6.5}$$

[Note that this formula is a rearrangement of the multiplication law: $p(A,B) = p(B) \times p(A \mid B)$, obtained by dividing both sides of the latter equation by $p(B)$.] As an illustration of the application of this formula we again consider the probability of being Eager, given an individual who is a woman, $p(A_2 \mid B_2)$.

We first find, by inspecting Table 6.2, the probability that the individual is both A_2 *and* B_2. This $p(A_2, B_2)$ value we find to be .30. Next we determine, from examining the column at the far right of the table, that the probability of B_2 is .60. This $p(A_2, B_2)$ value is then divided by $p(B_2)$, giving us .30/.60 or .50, the same figure for the conditional probability $p(A_2 \mid B_2)$ that we obtained earlier. What about $p(A_3 \mid B_1)$, the probability of being a *Dud*, given a *man*? The joint probability of A_3 and B_1, Table 6.2 shows, is .125, and the probability of B_1 is .40; the conditional probability of A_3 given B_1 therefore is .125/.40 or .3125. Application of this formula to all combinations of A | B (that is, probabilities of the three Enjoyment categories given the individual's sex) yields the conditional probabilities shown in Table 6.3. Note that for each B category the probabilities total 1.00; each man or each woman falls into one of the Enjoyment categories, so that the probabilities in the three Enjoyment groups must add up to unity in each sex.

Table 6.3 Conditional probabilities of the individual's being in each Enjoyment-Potential category, given the individual's sex, $p(A \mid B)$.

	Swingers (A_1)	Eagers (A_2)	Duds (A_3)	Total
Men (B_1)	.3125	.3750	.3125	1.00
Women (B_2)	.2083	.5000	.2917	1.00

Conditional probabilities can also be found for B, given A, by a parallel formula:

$$p(B \mid A) = \frac{p(A, B)}{p(A)} \tag{6.6}$$

For example, given an individual who is a Swinger (A_1), what is the probability that the individual is male (B_1)? From the joint probabilities in Table 6.2 we determine that the A_1B_1 probability is .125 and that $p(A_1)$ is .25; $p(B_1 \mid A_1)$ is therefore .125/.25 or .50. The complete set of $p(B \mid A)$ values is shown in Table 6.4. Again you will see that for a given class of A (Enjoyment

Table 6.4 Conditional probabilities of the individual's being a man or woman, given the individual's Enjoyment Potential, $p(B \mid A)$.

	Swingers (A_1)	Eagers (A_2)	Duds (A_3)
Men (B_1)	.50	.33	.417
Women (B_2)	.50	.67	.583
Total	1.00	1.00	1.000

Potential) the two categories of B (sex) account for all cases of that A; the two B probabilities therefore total 1.00 for each A.

RANDOM SAMPLES

Underlying our discussion of probability has been the assumption that the events we have observed have been *randomly sampled*. Indeed, from time to time we have referred to an individual or event as having been chosen "at random." We now can explain precisely what this expression means.

Consider our familiar standbys, a set of three balls labeled A, B, and C. These balls constitute a *population*—all the elements in a set of observations. We place them in a bowl and mix them thoroughly. Next, we select a ball and, without replacing it, select another. These two balls constitute a *sample*. If we write down the letters of the balls in the order in which they were drawn, we have six possible samples:

$$\begin{array}{ccc} AB & AC & BC \\ BA & CA & CB \end{array}$$

Assuming that we have shaken up the bowl thoroughly and picked blindly, each of the six samples has an equal chance of being chosen ($p = {}^1/_6$). We also notice that on the first draw each of the three balls has an equal chance of being chosen ($p = {}^1/_3$) and on the second draw each of the two remaining balls has an equal chance of being chosen ($p = {}^1/_2$).

We could also select a sample of two balls but *replace* the first before drawing the second. Here we have nine possible samples, having to add AA, BB, and CC to the six listed above. Again, each *sample* has an equal chance of being selected ($p = {}^1/_9$) and, since we are using a replacement procedure, each ball on each draw has an equal chance of being chosen ($p = {}^1/_3$).

Both of these examples are illustrations of random sampling, one with replacement and one without replacement. Random sampling, then, is a method of selecting a sample in such a way that each *member* of the pool has an equal probability of being selected on any given draw, and each possible *sample* of members has an equal probability of being selected.[2] The *result* of this selection procedure is known as a *random sample*. Many specific methods of selection will produce random samples. It is the equal likelihood of selecting each of the possible samples and each of the members of the population that defines randomness and not the particular procedure used to achieve this result. We emphasize that it is the individuals, not their scores, that must have equal likelihoods if sampling is to be called *random*.

PERMUTATIONS AND COMBINATIONS

A budding art collector has four paintings, which he displays on his living room walls. The wall hooks are fixed, but he regularly changes the places where each painting hangs. How many different spatial arrangements or orders of the four paintings are possible? This question concerns the number of different sequences or *permutations* a set of discrete objects can take. With two objects two orders or permutations are possible: AB and BA; with three objects, A, B, and C, six permutations are possible: ABC, ACB, BAC, BCA, CAB, and CBA. With four objects there are 24 permutations—an answer we could have figured out by jotting down and counting all the possibilities. We can determine the answer more easily by recognizing that the number of permutations (P) a set of N objects can take is equal to the product of the integers from N to 1. The product is called N *factorial* or N!. Thus:

$$P_N = N! = N(N - 1)(N - 2) \ldots (1) \qquad\qquad (6.7)$$

Application of this formula to three objects gives us $3 \times 2 \times 1 = 6$, and to four objects $4 \times 3 \times 2 \times 1 = 24$.

In the formula above you will note that P has a subscript, N. This

---

[2] In the case of a series of coin tosses our definition of random sampling is inapplicable, because we cannot specify a finite population from which we are drawing a sample. See Chapter 7 for predictions of the behavior expected in random coin tosses.

subscript indicates that we are trying to determine the number of possible permutations for N objects taken N at a time. But we may also be interested in N objects taken *r* at a time, where *r* is some number less than N. For example, our art collector may have increased his holdings to nine paintings but have room to display only five of them at one time. How many arrangements or permutations of these nine paintings, hung five at a time, are possible? The formula that permits us to answer this question is:

$$P_r^N = \frac{N!}{(N-r)!} \tag{6.8}$$

where the subscript *r* attached to P indicates that we are seeking the number of permutations for *r* objects at a time, rather than the entire set. On the right side of the equation the numerator is the same N factorial as above, while the denominator indicates that *r* is to be subtracted from N and the factorial of the resulting number found—that is, the product of the integers from (N − *r*) to 1. If six objects are taken two at a time, then the number of permutations is determined by $6!/(6-2)!$ or $(6 \times 5 \times 4 \times 3 \times 2 \times 1)/(4 \times 3 \times 2 \times 1)$. The result of these calculations is 30.

Our art collector with nine paintings that he wants to hang in all possible orders of five at a time faces an almost impossible task, since $9!/(9-5)!$ yields 15,120 permutations. He would have a better chance of succeeding if he contented himself with all possible *combinations* of his paintings. Combinations refer to the number of different subsets of *r* objects each that we can select from a set of N objects *without regard to sequence*. For example, three objects taken two at a time permit six sequential arrangements or permutations: AB and BA, AC and CA, BC and CB. If we disregard order (for example, treat AB and BA as the same), we see that there are three different subsets of objects; therefore the number of unordered subsets or *combinations* of three objects taken two at a time is 3. The formula for determining this number is as follows:

$$C_r^N = \frac{N!}{r!(N-r)!} \tag{6.9}$$

The number of ways that nine paintings can be displayed five at a time, ignoring the particular arrangement of a given subset of five on the walls, is therefore $9!/5!(9-5)!$. This equals 126, a more manageable number than 15,120.

Knowledge of permutations and combinations, along with the principles of probability, will help us understand binomial probability distributions and the normal probability distribution, discussed in the next chapter.

TERMS AND SYMBOLS TO REVIEW

Equally likely events

Mutually exclusive events

Set

Sampling with replacement

Sampling with nonreplacement

Addition rule $[p(\text{A or B})]$

Multiplication rule $[p(\text{A, B})]$

Independent events

Nonindependent events

Conditional probability $[p(\text{A} \mid \text{B})]$

Joint events

Probability table

Random sample

Permutation (P_N, P_r^N)

Combination (C_r^N)

N factorial (N!)

Gambler's Fallacy

The Binomial
and the Normal
Distributions

7

A textbook in statistics is a bit like a murder mystery. The chapters successively build up to the solution of an intellectual puzzle, a solution which typically becomes clear only near the end of the book. Along the way, the reader is often confused, misled, and taken down what seem to be false trails. However, unlike mystery story readers, statistics students are often unsure that there is any particular direction in which the chapters are leading them. Worse, authors of statistics texts do not purposely mystify their readers or hold off the denouement to create suspense; their intent is to the contrary. We therefore begin by putting up some signposts which point to where we are going before plunging into the thickets of the major topics to be discussed in the chapter.

Two Experiments

A wine-loving friend of yours claims his palate is so sensitive that he can discriminate between two wines of the same type, produced in the same year by neighboring vintners. He happens to have two bottles of wine that meet this description and you challenge him to prove his boast. On each of five "trials" you have him sip from each of two unmarked glasses and then announce which wine is which. On four of the five trials (on 80 percent of the occasions) his identification is correct. Should you be impressed by his claim and concede his ability to discriminate?

Consider a second experiment, in which groups of three subjects are shown 30 vertical lines, one after another. As each line is shown, the experimenter calls upon each subject, in order, to announce his estimate of its length. The first two "subjects" are actually confederates of the experimenter and systematically overestimate the true length.[1] By the end of the experimental series—let us say on the last ten lines—do the genuine subjects in these trios continue to report accurately? Or, under the pressure of their peers' judgments, do they begin to "see" the lines as longer—or perhaps even shorter—than they actually are?

Our first example illustrates a *binomial* experiment—a situation in which only one of two possible outcomes can occur on each of a number of independent occasions. The wine taster was given five independent opportunities to judge which of the two wines was which and on each occasion was either *right* or *wrong*. The second experiment involved *continuous* data.

---

[1] The experiment is modeled after a classic experiment by Solomon Asch (Asch, S. E. Studies of Independence and Submission to Group Pressure: I. A Minority of One Against a Unanimous Majority. *Psychological Monographs* Vol. 7, Series No. 416). Subjects were found to conform in their judgments to the confederates', a phenomenon that has come to be known as the Asch effect.

Spatial distance exists along a continuous, unbroken scale; thus, the length of lines presented to the subjects and hence the subjects' judgments about them could in principle take any value on the scale.

To answer the questions these experiments were designed to test, we should know something about two theoretical distributions: the binomial probability distribution and the normal probability distribution. This chapter will describe these distributions and their properties. In the next chapter we will consider how experimental questions such as those posed in our two examples may be formally evaluated.

THE BINOMIAL DISTRIBUTION

In the binomial situation one of two possible outcomes occurs on each occasion. Assume, for purposes of illustration, that each outcome has a probability of occurrence of $^1/_2$. For example, we might be tossing a fair coin in which H and T have an equal probability of turning up. Or the subject in our wine-tasting experiment might be unable to discriminate between the two wines and therefore be equally likely to be right as wrong. If we call the occurrence of a head or of a correct guess a "hit," and a tail or a wrong guess a "miss," we can determine by applying the multiplication rule the number of possible patterns of hits and misses on a given number of occasions and the probability of occurrence of each. If we toss a coin three times, for example, eight patterns of hits and misses are possible (HHH, HHM, HMM, HMH, MHH, MHM, MMH, MMM) and the probability of each is $(^1/_2)^3$. That is, the probability of each is $^1/_2 \times {}^1/_2 \times {}^1/_2$ or $^1/_8$. Similarly, with five occasions there are 32 possible outcomes, each having a probability of $(^1/_2)^5$ or $^1/_{32}$.

Now suppose that for the five-trial case we write out the 32 possible patterns of hits and misses. For each pattern we then count the number of *hits*, ranging from five out of five to zero. We find that five hits occur in only *one* of the 32 possible outcomes, but there are five patterns in which four hits occur (HHHHM, HHHMH, HHMHH, HMHHH, MHHHH), ten patterns in which three hits occur, and so on. A complete frequency distribution of the various numbers of hits, from five to zero, is shown in Table 7.1. In the final column of the table is the probability of occurrence of each number of hits. This list of probabilities is known as a *binomial probability distribution*.

The binomial probabilities in the table were obtained by applying the addition rule, discussed in Chapter 6. That is, each of the *individual* patterns of hits and misses has a probability of $^1/_{32}$ or .03125. Five hits occur only once among the 32 patterns, so its probability is .03125. Four hits occur in *five* of the patterns; adding .03125 together five times gives .15625, and so on.

Table 7.1   Theoretical frequency distribution of number of "hits," where number of occasions = 5 and $p = .50$.

Hits	Frequency	prob(Hits)
5	1	.03125
4	5	.15625
3	10	.31250
2	10	.31250
1	5	.15625
0	1	.03125
	$\sum f = 32$	$\sum prob(Hits) = 1.00000$

The Binomial Expansion

In explaining the binomial distribution above, we suggested writing down all the possible patterns of the two events that could occur on N occasions and then counting the number of patterns containing N of the specified ("favored") event, N − 1 of the favored event, and so forth down to N − N or 0. We can determine the distribution more efficiently by expanding the binomial:

$$(p + q)^N$$

where p is the probability of obtaining the favored event on a given occasion, q is the probability of the nonfavored event, and N is the number of independent occasions or observations. The expansion takes the general form:

$$(p + q)^N = p^N + Np^{N-1}q + \frac{N(N-1)}{1 \times 2} p^{N-2}q^2$$

$$+ \frac{N(N-1)(N-2)}{1 \times 2 \times 3} p^{N-3}q^3 + \cdots + q^N \qquad (7.1)$$

For $p = \frac{1}{2}$ and N = 5 we thus have:

$$\left(\frac{1}{2} + \frac{1}{2}\right)^5 = \left(\frac{1}{2}\right)^5 + 5 \left(\frac{1}{2}\right)^4\left(\frac{1}{2}\right) + 10 \left(\frac{1}{2}\right)^3\left(\frac{1}{2}\right)^2$$

$$+ 10 \left(\frac{1}{2}\right)^2\left(\frac{1}{2}\right)^3 + 5 \left(\frac{1}{2}\right)\left(\frac{1}{2}\right)^4 + \left(\frac{1}{2}\right)^5$$

Note the coefficients in each term in the expansion. The coefficient for the first term, $(^1/_2)^5$, is understood to be 1, and the remainder are 5, 10, 10, 5, and 1. These coefficients correspond to the frequencies in Table 7.1, while the terms themselves correspond to the probabilities listed in the table. For example, the probability of four hits is given by the second term, $5(^1/_2)^4(^1/_2)$; the value of this term equals .15625, the same figure we obtained earlier by the addition rule.

Each term in the binomial expansion can also be obtained by:

$$C_r^N p^r q^{N-r}$$

where C_r^N is the combination of N things taken r at a time. Since, according to formula (6.9), C_r^N is equal to $N!/r!(N - r)!$, we can write:

$$C_r^N p^r q^{N-r} = \frac{N!}{r!(N - r)!}\, p^r q^{N-r} \tag{7.2}$$

As an example of the application of this equation, we determine the probability of obtaining three hits in a sample of five observations, where $p = {}^1/_2$:

$$C_3^5 p^3 q^2 = \frac{5!}{3!(5 - 3)!} \left(\frac{1}{2}\right)^3 \left(\frac{1}{2}\right)^2 = (10)\left(\frac{1}{32}\right) = .3125$$

The coefficient (10) and the probability (.3125) agree, you will observe, with the term for three hits in the expansion of the binomial equation and the entries in Table 7.1.

Let us consider one more illustration of the application of the equation immediately above. If we rolled four dice, we could determine the probability of getting various numbers of, let us say, six-spots. Sixes could come up on all four dice, on only three of them, two of them, and so on. Since the probability of rolling a six-spot in a single die is $^1/_6$ and of rolling any other number $^5/_6$, we could determine the probability of these various numbers of six-spots in four rolls by expanding the binomial $(p + q)^N = (^1/_6 + {}^5/_6)^4$. What if we wanted to know only the probability of obtaining *two* sixes among the four dice? We could expand the binomial and use the appropriate term, or we could apply Formula (7.2). The latter procedure gives us:

$$C_2^4 p^2 q^2 = \frac{4!}{2!(4 - 2)!} \left(\frac{1}{6}\right)^2 \left(\frac{5}{6}\right)^2 = 6\left(\frac{25}{1296}\right) = .116$$

Before concluding this section, we return briefly to the results of the wine-tasting experiment with which we started. The would-be expert, you recall, succeeded in guessing correctly on four of five occasions. We can see from Table 7.1 that if he could *not discriminate at all* (that is, if the probability of a hit and a miss were equal), the probability that he would be this accurate *or more* by chance alone is about .19. (We obtained this figure by adding the probabilities for four hits and five hits.) His performance therefore is not particularly impressive, and on this evidence we probably would be quite skeptical about accepting his claim that he can discriminate. In the next chapter we will treat more formally the procedures used to reach decisions in such situations.

The Mean and Standard Deviation of the Binomial Distribution

It is possible to calculate the mean and standard deviation of a binomial distribution, just as we learned to do for any other kind of distribution. If you look at Table 7.1, for example, where $p = .50$ and $N = 5$, what is the mean number of hits? Since the distribution is perfectly symmetrical, the mean falls at the midpoint of the range, which we observe to be 2.5. We could also calculate the mean in the ordinary way by multiplying each hit by the corresponding p (which is essentially percent frequency), summing these products, and dividing by $\sum p$. ($\sum p$, of course, equals 1.) If you were to carry out these operations, you would find that the mean turns out to be 2.5, just as it should be. The formula for the mean of the binomial can be reduced to a simpler form that considerably lessens our calculational labors:

$$\mu_b = Np \tag{7.3}$$

where N is the number of observations and p the probability of the favored event. In our example μ_b therefore equals $5(.5)$ or 2.5. You will observe that we are dealing with a theoretical population of values, so the symbol we use for the mean is μ (mu). The subscript, b, indicates that we are referring to the mean of the binomial distribution.

We can determine the variance and the standard deviation of a binomial distribution in the usual way by finding the squared deviations of the "scores" from μ_b, finding the mean of these squared deviations, and so on. But just as in the case of μ_b, simpler formulas can be used. The *variance* of the binomial is given by:

$$\sigma_b^2 = Npq \tag{7.4}$$

and the *standard deviation* by:

$$\sigma_b = \sqrt{Npq} \tag{7.5}$$

where $q = 1 - p$.

Thus, in our example, we have for the standard deviation:

$$\sigma_b = \sqrt{(5)(.5)(.5)} = \sqrt{1.25} = 1.118$$

Shape of the Binominal Distribution When $p = {}^1/_2$

When $p = {}^1/_2$, the binomial distribution is always symmetrical. As N, the number of possible occurrences, becomes larger and larger, the shape of the binomial distribution with $p = {}^1/_2$ becomes increasingly similar to another theoretical distribution. This is the *normal probability distribution*, a particular member of the family of bell-shaped curves based on continuous rather than discrete measures. The similarity is illustrated in Figure 7.1, in which $p = {}^1/_2$ and N = 10. On the baseline we have indicated the number of "hits" (for example, number of heads coming up in a toss of ten coins). On the Y axis is shown the frequency, out of the 1024 possible patterns of hits and misses, with which each *number* of hits could be expected to occur. (These frequencies,

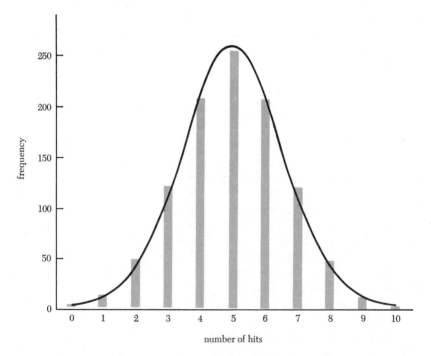

Figure 7.1 A binomial frequency distribution in which $p = .50$ and N = 10 and a superimposed normal distribution with the same mean and SD, $\mu = Np = 5$, and $\sigma = \sqrt{Npq} = \sqrt{2.5}$.

you recall, correspond to the coefficients of the terms in the binomial expansion.) The frequencies have been plotted as bars to emphasize that the measures are discrete. That is, they take the values 0, 1, 2, 3, and so on; intermediate values such as 1.54 hits or 3.791 do not occur. Superimposed on the bar graph for the binomial distribution is a normal probability curve whose mean and SD have the same values as those of the binomial. Since continuous measures (those that can take any value) are represented, the curve is unbroken.

You will observe that the two shapes are highly similar. As the N in a binomial increases toward infinity, the size of the "steps" between adjacent bars becomes smaller and smaller, and the distribution increasingly resembles the continuous normal distribution. Thus, we can state that the *limit of the binomial distribution*, where $p = {}^1/_2$ and N is infinitely large, is the *normal probability distribution*.

The similarity between the two distributions often leads us to substitute the normal distribution for the binomial distribution when we are evaluating binomial data like those in our wine-tasting experiment. When $p = {}^1/_2$, this substitution may safely be made when N is relatively small (as low as 10, let us say). As p departs from ${}^1/_2$ (toward 0 or 1), the normal curve may provide a satisfactory approximation but N must be increasingly larger for the normal curve to be used. We shall discuss this use of the normal curve more specifically in the next chapter. We will devote the rest of the present chapter to describing the properties of the normal curve.

THE NORMAL PROBABILITY DISTRIBUTION

Like the binomial distribution, the normal probability distribution is a theoretical mathematical model, based on a population of measures that is infinite in size. But there is an infinite variety of curves, and by choosing the appropriate mathematical equation we can plot and describe the properties of any of them we wish. Why, then, are we giving so much emphasis to this particular curve? The primary reason has to do with statistical theory. We have already seen that the binomial distribution approaches this form as N increases, and we shall see in later chapters that other theoretical sampling distributions also take this form.

A second reason why the normal curve is of interest is that many empirical distributions—those obtained by actually measuring some characteristic—are approximately normal in shape. That is, if we attempt to fit an idealized mathematical shape to the data we obtain, the equation for the normal distribution is an appropriate model. By applying our knowledge of the properties of the normal curve to an empirical distribution, we can answer many types of questions about the distribution.

In explaining the properties of normal curves we will use only empirical distributions for illustrative purposes, deferring discussion of the curve's theoretical uses until later chapters.

Characteristics of the Normal Curve

The normal curve is shown in Figure 7.2. Since it is a plot of a theoretical distribution and not of actual data, the units of measurement along the baseline are expressed in SD units (z scores). We should also recall that the distribution is based on an infinite number of cases, which can take any z-score value. The curve therefore is not dropped to the baseline at either end and can be understood to reach it only at z's of $\pm\infty$.

The most obvious characteristic of the normal curve is its shape, somewhat like a bell, rising to a rounded peak in the middle and tapering off symmetrically at both tails. The half of the curve below μ has an $\diagup$ shape and the half above μ a reversed $\diagup$. The vertical lines erected from the baseline at distances of $+1\sigma$ and -1σ both intersect the curve at its "inflection point"—the point at which the $\diagup$ and its mirror image each change directions, from being concave to being convex.

The total *area* under the curve—the area between the curve and the baseline—is taken as 100 percent. We can also mark off portions of the curve and describe their areas as a percent of the total. In Figure 7.2, for example, we have dropped lines from the curve to the baseline at μ and at one SD above μ. Now in *any* curve, normal or any other shape, the portion of the total area under the curve between μ and the point that is one SD above μ is always the same for *that particular curve*. In other words, for any single curve, regardless of its shape, there is a fixed relationship between distance along the baseline measured in SD units and area under the curve. In the

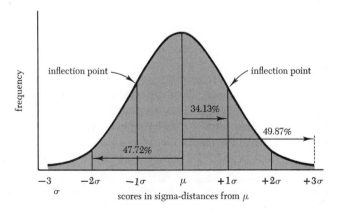

Figure 7.2 The normal distribution curve

normal curve the area under the curve between μ and $+1$ SD turns out to be 34.13 percent of the total area. *The percent area is the same as the percent frequency*; therefore, we know immediately that in the normal curve 34.13 percent of the total frequency is included between μ and the point that is one SD above μ. *When distance along the baseline is measured in SD units, percent area under the normal curve between μ and any other point is the same as percent frequency.* Recall that SD units are z scores; the score that is one SD above the mean is the score where $z = +1.00$. Thus, we can restate what we have just said in z-score terms; in the normal curve the percent frequency between μ and the point where $z = +1.00$ is 34.13.

Since the normal curve is symmetrical, another 34.13 percent of the frequency lies between μ and the point one SD below μ, between μ and $z = -1.00$. Therefore, 68.26 percent of the cases fall between $z = -1.00$ and $z = +1.00$. From μ to the point two SD's from μ is 47.72 percent of the frequency, so 95.44 percent fall between $z = -2.00$ and $z = +2.00$. Between μ and three SD is 49.87 percent. Thus, three SD's above and below μ, six SD's in all, include 99.74 percent, or almost all, of the total frequency.

Let us illustrate what we have stated. If the weight distribution for a sample of 10,000 men were exactly normal, 3,413 men (34.13 percent of N) would fall between the mean weight and the weight that was one SD above the mean. If the mean were 170 and SD 15, then 68.26 percent or 6,826 of them could be expected to weigh between 155 and 185 (170 $\pm$ 15, the scores where $z = \pm 1.00$).

As a final illustration, consider the distribution of measurements of chest circumference of American women. Since all women have not been measured, the actual μ and SD are not known. But the results of several studies[2] permit us to estimate that μ of this distribution is about 35 inches, with SD $= 3$ inches. Probably the distribution is sufficiently normal to permit us to predict that about 95 percent of American women would achieve, naturally, scores between 29 and 41 inches ($\mu - 2$ SD's and $\mu + 2$ SD's). Almost all women would score between 26 inches (35 $-$ 9) and 44 inches (35 $+$ 9)—that is, between $z = -3.00$ and $z = +3.00$.

Probability and Percent Area

We have discussed the percent of the total area in a normal curve that lies between representative points on the baseline, and the relationship between these percent areas and percent frequencies. For example, approximately 34 percent of the area, and hence about 34 percent of the total number

---

[2] *Handbook of Human Engineering Data for Design Engineers.* Tufts College Institute for Applied Experimental Psychology. Special Devices Center, Technical Report No. SDC 199-1-1, 1949.

of cases, falls in the interval between μ and one SD above μ. These area values also reflect *probabilities*. For example, the probability that an individual selected at random from a normal distribution will fall between μ and one SD above μ (will have a z score ranging from 0 to $+1.00$) is .34.

When we were discussing equally likely discrete events in the last chapter, such as the outcome of coin tossing or drawing cards from a deck, we defined probability as the ratio of the number of times the specified event can occur to the total number of possible events. The normal curve, however, involves continuous rather than discrete variables. Therefore, we must employ a somewhat different definition of probability for events randomly sampled from a normal distribution. This definition states that probability in the normal curve is the *ratio of the area under the specified portion of the curve to the total area*. This definition, you will observe, states formally what we have said in the paragraph above.

The Normal Curve Table

The percent area under the normal curve, and therefore the percent frequency, between μ and various SD distances from μ is shown in Table B at the end of the book. This table will be used repeatedly from now on. Let us examine the table and determine how to read it. The left-hand column, labeled x/σ, *gives deviations from μ in SD units*. Remember that x stands for the deviation of a score from the mean of the distribution containing the score: $x = X - \mu$. Dividing the deviation, x, by the SD of the distribution, σ, tells us how far the score deviates from μ in SD units. But $(X - \mu)/\sigma$ is the definition of z; *therefore, the values in the left-hand column in Table B are z scores*. An x/σ value, or z score, of 1.00 indicates a raw score that deviates one SD from its mean; an x/σ value of 2.58 indicates a raw score that deviates 2.58 SD's from μ, and so forth.

The values in the x/σ column in Table B are given to one decimal place; the second decimal place is shown in the top row of the table. The values in the body of the table are the *percent area* or *percent frequency* between μ and the point x/σ distant from μ. For example, if we wished to determine the percentage of the total frequency between μ and the point 1.96 SD's above μ, we would first find the value of 1.9 in the x/σ column, then go along that row to the column headed .06 and read off the percent frequency, which in this case is 47.50 percent.

Note that the table gives percentages only on *one* side of μ—that is, for only one-half of the normal curve—so none of the values in the table is larger than 50 percent. Because the normal curve is symmetrical, it is not necessary to print another table of the percentages of the other half of the curve; the percent frequency between μ and any x/σ value is the same whether the x/σ value is above or below μ, whether the z score is plus or minus.

Using the Normal Curve Table to Solve Problems

Let us now take up some problems involving real data and show how the normal curve table can be used to solve them. To avoid terminological confusion, we will assume that each group that has been measured constitutes a population of some sort, even if only a highly artificial one. Therefore, we will assume that the means and SD's of the obtained distributions are population values and refer to them, respectively, by the symbols μ and σ.

It may be helpful to state our general procedure first. If our empirical distribution is normal, or approximately so, the relation shown in Table B between percent frequency and z scores in the normal curve also holds for the empirical distribution. But the raw scores in an empirical distribution may be in any kind of units, such as feet, seconds, letters cancelled per four minutes, and the like. In the normal curve the units are z scores, x/σ values. Therefore, we shall have to convert the raw score from the empirical distribution, given us in the problem, into a z score or x/σ value, and then immediately Table B becomes applicable. Let us see how this works.

In a study[3] dealing with various athletic skills, 203 men students at the University of California were measured as to how far they threw a baseball. The mean was 164.1 feet, the SD was 22.8 feet, and the distribution was approximately normal. With only this information we could answer some questions immediately; for example, we know that the average (mean) student should have no trouble throwing from third base to home plate (90 feet). But consider the following questions. How many men in this group could throw to home plate from a spot 200 feet away in center field? What proportion of the men could not throw farther than 130 feet? Anyone who qualifies among the top 10 percent of the men must be able to throw at least how far? What is the probability that someone picked at random from this group could throw 225 feet or more? What distances are so extreme that they are thrown by only 1 percent of the men? What extreme distances are thrown by only 5 percent?

Since it will not affect the method we are illustrating, we shall make the data more convenient to handle by rounding off μ to 164 feet and SD to 23 feet, although we would not do this in actual practice. We also will assume for purposes of illustration, that the data were normally distributed. The *first question* we asked concerning this distribution was, how many men in this group could throw to home plate from a spot 200 feet away in center field? This problem simply asks how many men (a frequency) could throw 200 feet *or more* (because anyone who can throw farther than 200 feet can

---

[3] Cozens, F. W. The measurement of general athletic ability in college men, *Physical Education Series*, No. 3, University of Oregon Press, 1929.

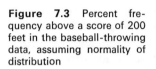

Figure 7.3 Percent frequency above a score of 200 feet in the baseball-throwing data, assuming normality of distribution

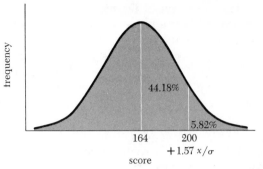

also throw 200 feet). This problem is shown graphically in Figure 7.3; it will help to study this figure in connection with the following. The raw score of 200 feet must first be converted to an x/σ value: $x/\sigma = (200 - 164)/23 = +1.57$. (We use the $+$ sign to indicate that the raw score with which we are dealing is above μ.) From Table B, 44.18 percent of the frequency are included *between* μ and the x/σ value $+1.57$, therefore 5.82 percent ($50 - 44.18$) remain *beyond* $+1.57$. We now have the answer to the problem in percent frequency, but we were asked for the actual number of men, the raw frequency. Since there were 203 men, 5.82 percent, or about 12 men, can throw 200 feet or more.

The second question asked was, What proportion of the men could not throw farther than 130 feet? Again, note carefully what this question asks for: the proportion of men whose longest throw was 130 feet or *less*—in other words, all those who cannot throw more than 130 feet. Converting the raw score to a z score, $(130 - 164)/23$, we get $x/\sigma = -1.48$. From Table B we find that between μ and $x/\sigma = -1.48$ are 43.06 percent, so 6.94 percent remain beyond this x/σ value. This is the answer in percentage; to convert to proportion move the decimal point two places to the left; thus, .0694 is the proportion of men who could not throw farther than 130 feet.

The third question was, Anyone who qualifies among the top 10 percent of the men must be able to throw at least how far? Note the difference between this question and the previous ones. Instead of being given a raw score and ending up with a percent frequency or a raw frequency, here we are given a percent frequency and asked for a raw score. Thus we shall go through the same steps as before but in the opposite direction. We first must find, from Table B, the z score, or x/σ value, *beyond* which 10 percent of the cases fall, because the problem asks for the top 10 percent. Remember that Table B gives frequencies *from* μ *to* a particular x/σ; therefore, we must find in the body of the table the value 40 percent; this will give us the z score beyond which 10 percent remain. The closest value to 40 percent we can find in Table B is 39.97 percent, which corresponds to $x/\sigma = +1.28$. We now know that the top 10 percent of the men deviate at least 1.28 SD's above μ, but the problem asks for a raw score, so we must convert the z score of 1.28 into a raw score. Recall that we can transpose terms in the z score formula ($z = (X - \mu)/\sigma$) to get $X = z\sigma + \mu$. Since SD is 23 feet, these men deviate 29 feet from μ ($1.28 \times 23 = 29.44$). Adding this value to μ (164 feet), we find that the top 10 percent of the men can throw at least 193 feet. Note again that the steps in this problem are the same as in the previous two problems, but we went through them in reverse order.

The fourth question asked: What is the probability that someone chosen at random from the group of men could throw 225 feet or more? This is the same as asking how frequently scores of 225 feet or more occur. First, as always, the raw score must be converted into a z score before Table B can be used: $(225 - 164)/23 = +2.65$, the x/σ value. Note that this problem asks for the probability of a score at least 2.65 SD's *above* μ; that is, the direction of the deviation is specified. By means of Table B we determine that 49.6 percent of the cases are between μ and $x/\sigma = 2.65$; therefore, .4 percent deviate more than 2.65 SD's. Thus, the probability that someone could throw 225 feet or more is .4 in 100, 4 in 1000, 1 in 250. Therefore p is equal to .004.

The fifth question, What distances are so extreme that they are thrown by only 1 percent of the men, also gives us a percent frequency and asks us to find a score. But, unlike the previous questions, this problem does *not* tell us whether high scores or low scores are desired. Therefore, we will assume that the 1 percent means the top one-half percent plus the bottom one-half percent. Note that in the previous three problems the direction of the deviation from μ was implied in the question. For example, we asked how many men could throw 200 feet or more. We therefore asked *only* for those who deviated at least 1.57 SD's *above* μ. Perhaps all this will be clearer if we restate the problem we are working on as follows: what scores (distances of throw) deviate from μ to such an extent that the *total* frequency of scores remaining beyond the deviation is 1 percent? We now can see more easily that we are asking for the top one-half percent and the bottom one-half percent of the scores.

The problem is shown graphically in Figure 7.4. The raw-score values given in the figure as the answer to the problem are found by our usual procedure. First find in Table B the x/σ value beyond which only one-half percent of the frequency remain, the x/σ value that includes 49.5 percent between it and μ. In the table the value 49.5 percent falls halfway between the x/σ values 2.57 and 2.58, so we can choose either one. We shall arbitrarily decide on 2.58, and we now have the answer to the problem in z-score terms: scores that occur 1 percent of the time are all those that deviate at least 2.58 SD's from μ, including both directions from μ. However, we were asked for the raw scores, so we must convert the z score of 2.58 to raw-score units.

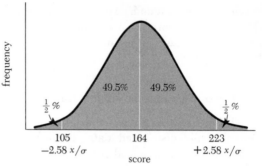

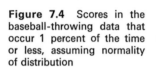

Figure 7.4 Scores in the baseball-throwing data that occur 1 percent of the time or less, assuming normality of distribution

Since SD of the distribution is 23 feet, 2.58 SD's are 59 feet (2.58 × 23 = 59.34). Now we have the answer as a deviation in raw-score units: distances in the baseball-throwing data that occur 1 percent of the time are all those that deviate at least 59 feet from μ in either direction. Since μ is 164 feet, these are distances of 105 feet or less and 223 feet or more.

The problem we have just solved was chosen to illustrate something we shall now emphasize. The problem asked for the deviation beyond which a certain percent frequency occurs without specifying the direction of the deviation. As we shall find in later chapters, this type of problem occurs often. We may state as a general rule that whenever the direction is *not* specified in a problem, you always assume the deviation in both directions is meant. In other words, when you are given a percent frequency and have to find the deviation corresponding to it, *unless it specifically states otherwise* the problem means that the deviation be such that half the percent frequency remains beyond it at each end of the curve.

Another reason why the above problem was chosen is that in a later chapter, when we take up the important problem of testing hypotheses, we shall find that the frequency of 1 percent, one-half percent under each tail of the curve, has special significance. You should memorize the x/σ value corresponding to the 1 percent frequency. *Remember* that scores that deviate at least 2.58 SD's from μ, considering both directions from μ, occur 1 percent of the time.

The *final question* asked the distances that could be thrown by the extreme 5 percent of the men. This question is of the same type as the one we have just answered. It inquires about both extremes of the distribution, 2.5 percent at one end and 2.5 percent at the other. We must look in Table B for the deviation that includes 47.50 percent between it and σ; doing so, we find the x/σ value to be 1.96. We now convert the z score of 1.96 into raw-score units. Since SD = 23, 1.96 SD units is equal to a raw score of approximately 45 feet (1.96 × 23 = 45.08). The extreme 5 percent can throw 119 (164 − 45) feet or less or 209 (164 + 45) feet or more. The frequency of 5 percent—2.5 percent under each tail of the curve—also has special significance when we test hypotheses about populations. You should memorize that 5 percent of the cases in a normal distribution equal or exceed z scores of ± 1.96.

Percentiles from the Normal Curve

Table B is also useful for determining percentile ranks for raw scores from normally distributed data. It must be emphasized, however, that percentile ranks determined from Table B will not be accurate unless the raw scores are distributed normally or very nearly so. Recall that the percentile rank of a given score is the percentage of the total number of cases, the percent frequency, lying *below* that score. To illustrate, assume a normal distribution with a mean of 121 and SD of 20. What is the percentile rank of the raw score 141? As usual, we have to get the z score corresponding to the raw score; in this case $z = +1.00$. From Table B, 34.13 percent fall between this x/σ value and μ. Therefore, the raw score 141 is above this 34.13 percent of the cases plus the 50 percent that are below μ, so the percentile rank is 84 (rounded). As you can see, we have merely found the total percent frequency falling below the raw score given. In other words, to use Table B to find the percentile rank for a raw score from normally distributed data, we have only to perform the usual step of converting the raw score to an x/σ value and then, using Table B, find the total percent frequency falling below the x/σ value.

Let us also illustrate how to find the raw score corresponding to a known percentile rank; for example, what score in the distribution above (where $\mu = 121$ and $\sigma = 20$) corresponds to the 31st percentile? This must be the score below which 31 percent of the cases fall, which means 19 percent of the cases are between that score and μ. From Table B we find that .50 is the x/σ value most closely corresponding to 19 percent. We now know that the raw score we are looking for is $^1/_2$ SD below μ; therefore the 31st percentile corresponds to a score of 111.

We have now considered a variety of normal curve problems; before going further it may be helpful to point out how similar they are. We should find the solution of normal curve problems easier if we keep in mind that all problems will give us some information expressed in one kind of unit, and that to answer the problem the given information will have to be translated into some other kind of unit. Specifically, the problem will give us at least one of the following units: a raw score, a z score or x/σ value, a percent frequency, or a raw frequency. Answering the problem will require translating from one to another of these units. Perhaps the diagram below will make this clearer. The top row shows the general procedure in all problems: we are given one kind of unit and we translate into another kind of unit. The two-headed arrows indicate that we can translate in either direction. Note, however, that whether we are translating from left to right or right to left, *we cannot skip a step*. To get from X to %f we have to go from X to z, then from z to %f; to get from f to X, we must first convert f to %f, then %f to z, then z to X.

$$X \longleftrightarrow z \longleftrightarrow \%f \longleftrightarrow f$$

$$\frac{X - \mu}{\sigma} = z \longrightarrow \text{Table B} \longrightarrow \frac{\%f(N)}{100} = f$$

$$X = z\sigma + \mu \longleftarrow \text{Table B} \longleftarrow \%f = \frac{f}{N}(100)$$

The other two rows of the diagram show the specific steps to take. Thus, suppose a problem gives a raw score (X) and asks for the percent frequency (%f), or raw frequency (f), above or below that score. For example, how many men scored below 150 feet in the baseball-throwing distribution? We would proceed from *left to right* in the top row of the diagram and we would go through the steps shown in the *second* row.

On the other hand, suppose the problem gives a percent frequency or raw frequency and asks for either a z score or raw score. As an example, the top 30 men are above what score in the baseball distribution? In this case we would go from *right to left* in the top row of the diagram and would go through the steps shown in the *bottom* row.

The point we want to make is that there are really not several different kinds of normal curve problems. All normal curve problems are essentially the same; they only look different because they can be stated in so many different ways. One more thing: we strongly advise drawing a graph for each problem. You often will find that if you express graphically the information given in a problem, it will help you get started answering it.

TERMS AND SYMBOLS TO REVIEW

Binomial probability distribution

SD of binomial (σ_b)

Binomial expansion

Normal probability distribution

Mean of binomial (μ_b)

Inflection point

Sampling
Distributions
Single Samples

8

In Chapter 7 we discussed at length the binomial distribution—the distribution expected from a large number of samples of events in situations in which each event has one of two possible outcomes. For example, in Figure 7.1 we plotted the frequency, in an infinite number of samples of 10 coin tosses each, with which samples with 0 to 10 heads would theoretically occur. Although we did not then describe it as such, Figure 7.1 represents what is called a *sampling distribution*. A sampling distribution is a theoretical distribution, based on a mathematical model, that represents the *distribution of values of a statistic that would be obtained from an infinite number of random samples of a given size.*

In this chapter we will extend our discussion of sampling distributions to include the distribution of sample *means*, based on the measurement of continuous variables, and will consider the use of sampling distributions in making inferences about population characteristics. Our first step will be to review the distinction between a sample and a population.

POPULATIONS AND SAMPLES

A population consists of all members of a group of individuals who are alike on at least one specified characteristic. (Populations may consist of any conceivable kind of animate or inanimate objects or events, but for illustrative purposes we will most commonly refer to people.) A population can be infinite or indefinitely large in number, such as all the infants who have ever been or ever will be born, or it can be finite, such as all the infants born last year in the United States. Finite populations may be large, such as all individuals enrolled in college, or they may be small, such as all majors in Greek at a given college. Populations may be real, as in the examples immediately above, or they may be hypothetical. If, for example, we were to conduct an experiment using a new drug designed to overcome psychotic depression, we would be concerned with estimating the drug's effect on the population of *all* individuals, now or in the future, who might exhibit this condition and *could* be given the drug.

Whether real or hypothetical, the population is specified by the investigator, and the aim is to find out something about it. The populations that are the subject of scientific investigation typically are large or indefinite in number, so that only a sample of the members can be studied. *A sample is any number of cases less than the total number of cases in the population from which is it drawn.* If we have a population of 50,000, a sample might consist of 10 cases, 15,000 cases, or 49,999. Normally, however, samples consist of only a small or a very small proportion of the population being investigated.

The investigator's purpose in observing a sample, we have said, is to use the data he obtains to make inferences about the characteristics of the population as a whole. (He may, of course, want to study and compare the characteristics of samples from more than one population in a single study, such as the value systems of individuals from different countries, arithmetic skills of elementary school students taught by several different methods, and so on.) It has probably already occurred to you that there are certain dangers in generalizing from sample to population data. If we were attempting to predict the outcome of a city mayoral election, for example, we would be foolish to question voters from only one occupational group or income level, since these factors are likely to be related to voter preference. What we would attempt to get is a sample of voters that is a miniature picture of the population from which it is taken. If we want to reach conclusions about a population from the data of a sample, the sample ought to be *representative* of the total population from which it is drawn.

We often try to obtain a representative group of individuals by drawing (measuring) a *random sample* from the population. A random sample, we have said, is a sample in which every individual in the population and every possible sample of individuals have an equal probability of being chosen. We have no guarantee that the measures obtained from a random sample will accurately reflect the properties of the population as a whole, but statistical theory does allow us to make probability statements about how closely data obtained from random samples can be expected to agree with population values.

VARIABILITY OF MEANS

If we measure the amount of some characteristic that members of a sample exhibit, the first statistic we would probably compute from the sample data is $\overline{X}$. The likelihood is small that this $\overline{X}$ has *exactly* the same value as μ of the population from which the sample was selected. That is, if we measured a large number of samples from the population, we would discover that the values of the sample $\overline{X}$'s were varied and that few were exactly the same as μ. If we were to use any given $\overline{X}$ as an estimate of μ, then, our estimate would be likely to contain a certain degree of inaccuracy. This fact is known as the *sampling error* of $\overline{X}$'s with respect to μ.

In Chapter 5 we said that the value of $\overline{X}$ was the best *single* guess we could make about the value of μ. But if $\overline{X}$ is not expected to be the same as μ, how can we use the sample $\overline{X}$ as indicative of the population mean? The best answer is, first, that sample means (unlike SD's) show no systematic tendency to be either larger or smaller than μ and hence can serve as an unbiased estimate of μ. Second, the mean of a truly random sample rarely deviates far from the population mean, particularly if the sample is large, so that $\overline{X}$ is usually a fairly close approximation of μ. In practice, however, we seldom use $\overline{X}$ to assign a precise value to μ; instead we use the sample data to test some specific hypothesis about the value of μ or to make estimates about a range of values that might reasonably be expected to contain μ. We will go on now with further elaboration of this important problem.

MEANS OF SUCCESSIVE RANDOM SAMPLES

Strictly speaking, the steps in reasoning we are going through now are based upon the assumption that we are drawing an infinite number of samples from an infinite population. However, it is quite satisfactory for illustrative purposes to go through the statistical logic with a finite population and a finite number of samples. We shall follow this procedure for the clarity it affords.

Suppose, then, we take successive random samples from a large, known population. We draw one sample of subjects and measure them on some aspect of behavior; then we replace the sample and draw another of the same size, measure the subjects in it, and so on. Let us say we draw 5,000 samples, calculate $\overline{X}$ for each, and make a frequency distribution of these $\overline{X}$'s. Such a distribution is called the *sampling distribution* of $\overline{X}$. What is the nature of the distribution of these $\overline{X}$'s? If the samples are relatively large, we find that if we group the $\overline{X}$'s and place them along the baseline (X axis) with frequency along the ordinate (Y axis), they approximate a *normal distribution*, even if the distribution of measures in the population as a whole is not normal. As the size of the individual samples is increased, the distribution of $\overline{X}$'s approaches the normal curve more and more closely, and the $\overline{X}$'s of the samples are increasingly likely to be close to μ. These relationships comprise the _central-limit theorem_, whose implications for making inferences about population values from sample data we will explore shortly.

Suppose that the 5,000 random samples we have obtained each have an N of 100 and are drawn from a population whose μ and σ we happen to know. Let us say that these values are 90 and 20, respectively. The distribution of the 5,000 samples will be (approximately) normal, as shown in Figure 8.1. What do we expect the mean of this distribution of 5,000 sample $\overline{X}$'s to be? Recall that $\overline{X}$'s are unbiased estimates of μ. That is, overall we do not expect the $\overline{X}$'s to be systematically larger or smaller than μ. Further, we expect errors of overestimation of a given size to occur equally often as errors of underestimation of μ. The mean of a sampling distribution of $\overline{X}$'s, therefore,

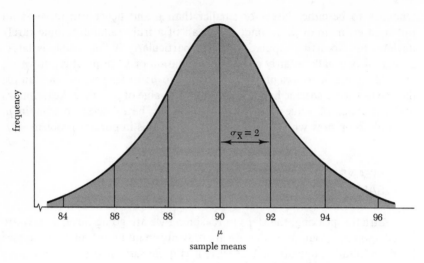

Figure 8.1 Sampling distribution of $\overline{X}$'s of 100 cases each from a population in which $\mu = 90$ and $\sigma = 20$

theoretically can be expected to take the same value as μ for the population. We have said that our μ is 90; thus, in Figure 8.1 the mean of the sampling distributions of $\overline{X}$'s takes this value.

In Chapter 7 we learned that we could make certain probability statements about where a raw score would fall in a normal distribution. For example, we know that an individual drawn at random from a normal population has a probability of approximately .68 of falling between the points established by $+1\sigma$ or -1σ. Similarly, we know that an individual has a probability of .05 of falling at or beyond the point established by $+1.65\sigma$. Since for all practical purposes our sampling distribution of $\overline{X}$'s can be assumed to be normal, we can make similar probability statements about a single sample $\overline{X}$. Thus, for example, the probability that an $\overline{X}$ will fall within the interval bounded by one SD above and one SD below the mean of the sampling distribution is .68.

We now ask a more concrete question about the distribution in Figure 8.1. What is the probability that an $\overline{X}$ will deviate from the μ of 90, in either direction, by 2 or more points? That is, what is the probability of obtaining $\overline{X}$'s of 92 or greater or 88 or less? If we knew the SD of the sampling distribution, we could answer the question. This SD, we now note, is given a special name, the *standard error of the mean*, and is symbolized by $\sigma_{\overline{X}}$. The word "error" is used in labeling this particular SD because each $\overline{X}$ in the distribution may incorporate a sampling error in the sense that each $\overline{X}$ will usually differ from the mean of the population.

We could, of course, calculate the SD of the sampling distribution ($\sigma_{\bar{x}}$) by treating each $\bar{X}$ as if it were a raw score and applying the formulas we learned in Chapter 5. However, an alternate method of calculation is available. Statisticians have demonstrated that the value of the standard error of the mean ($\sigma_{\bar{x}}$) has a constant relationship to the SD (σ) of the population from which the samples were drawn and the size (N) of each of these samples. The exact relationship is given by the formula:

$$\sigma_{\bar{X}} = \frac{\sigma}{\sqrt{N}} \qquad (8.1)$$

We specified earlier that the σ of our population is known to be 20. Since each sample N is 100, $\sigma_{\bar{x}}$ can be calculated to be:

$$\sigma_{\bar{X}} = \frac{20}{\sqrt{100}} = 2$$

The sampling distribution in Figure 8.1 thus has $\mu = 90$ and $\sigma_{\bar{x}} = 2$. We now return to our question: how probable is it that a sample $\bar{X}$ will deviate from μ, in either direction, by 2 or more points? We first must convert this deviation into a z score, defined, you recall, as the score minus the mean of the distribution, divided by its SD:

$$z = \frac{\bar{X} - \mu}{\sigma_{\bar{X}}} = \frac{x}{\sigma_{\bar{X}}} \qquad (8.2)$$

The "score" in this instance is $\bar{X}$, the mean of the sampling distribution is μ, and the distribution's SD is $\sigma_{\bar{x}}$. Thus $\pm 2/2$ is equal to ± 1.00. Since approximately 68 percent of the area in a normal curve lies between z's of $+1$ and -1, 32 percent lies *beyond* these two points (16 percent at either end). The probability of drawing a sample whose $\bar{X}$ is 92 or greater or 88 or less therefore is .32.

Our aim, however, is not to learn how to make probability statements about obtaining sample $\bar{X}$'s of certain values, given the values of the population μ and SD. If we had this information about the population, there would be little purpose in measuring samples. Our goal is the reverse: to make inferences about the unknown μ, given data from a single random sample. If we knew the SD of the sampling distribution of $\bar{X}$'s, we would be able to make these inferences about μ. This would not seem to put us very far ahead, since we do not know either the SD of the population (σ) or the SD of the sampling distribution (standard error of the mean or $\sigma_{\bar{x}}$). We can, however, calculate an *estimate* of $\sigma_{\bar{x}}$ from the data of a sample.

Estimate of Standard Error of the Mean

The $\sigma_{\overline{X}}$ of a sampling distribution can be determined, we have seen, by dividing σ, the SD of the population, by the square root of N, the number in each sample. It may occur to you that we could use this formula as a basis for estimating the standard error of the mean from the data of a single sample by using the SD of our sample, $\tilde{\sigma}$, as an approximation of σ and substituting it in the formula. This is a good start, but there is a problem. As you will remember from our discussion in Chapter 5, the SD of a sample tends to be smaller than the SD of its population. However, we can compensate for this underestimation by dividing $\tilde{\sigma}$ by the square root of N $-$ 1 instead of N. Thus, our *estimate* of the standard error of the mean (which we will identify by the symbol $s_{\overline{X}}$) is found by the formula:

$$s_{\overline{X}} = \frac{\tilde{\sigma}}{\sqrt{N - 1}} \qquad (8.3)$$

All we have to do to find $s_{\overline{X}}$ is to find the SD of the sample scores and divide this SD by the square root of one less than the number of cases in the sample.

An alternate formula that gives the same answer as Formula (8.3) is:

$$s_{\overline{X}} = \frac{s}{\sqrt{N}} \qquad (8.4)$$

where s is the estimate of the population SD (σ) derived from our sample data. We stated in Chapter 5 that s is obtained by taking the square root of $\sum x^2/(N - 1)$. Thus, in essence, the difference between the two ways of computing $s_{\overline{X}}$ is the place at which the correction factor of -1 is introduced to compensate for the fact that sample SD's tend to be smaller than the SD of their population.

Before proceeding further, let us state once more what this estimate of the standard error of the mean ($s_{\overline{X}}$) represents. It is an estimate, based on the data from a single sample, of the *standard deviation of a distribution of randomly drawn sample means*. In other words, $s_{\overline{X}}$ is an estimate of the variability of the sampling distribution of $\overline{X}$'s.

We now can demonstrate the calculation of $s_{\overline{X}}$ from sample data. Suppose we have a sample of 145 scores with $\tilde{\sigma}$ of 10.82. To find $s_{\overline{X}}$ we simply substitute in the formula:

$$s_{\overline{X}} = \frac{\tilde{\sigma}}{\sqrt{N - 1}} = \frac{10.82}{\sqrt{145 - 1}} = \frac{10.82}{12} = .90$$

The s calculated from the sample data turns out to be 10.86. Therefore, $s_{\bar{X}}$ is also given by:

$$s_{\bar{X}} = \frac{s}{\sqrt{N}} = \frac{10.86}{\sqrt{145}} = .90$$

We now have before us the statistical logic that will permit us to make inferences about population μ's from sample data. Two possible approaches are open to us. On occasions we wish to test a hypothesis that the μ of a population is some particular value and collect data from a sample of individuals so that we may evaluate how likely it is that the hypothesis is true. On other occasions we have no prior hypothesis about the value of μ but instead want to find a range of values within which μ probably falls.

TESTING HYPOTHESES ABOUT μ

We conduct a study to determine the beliefs of American middle-class adults about the reasons for poverty. One of the questions asks for a rating, on a 7-point scale, of the relative interest of the poor and nonpoor in having material things. A scale point of 1 indicates that the poor are very much *less* materialistic than the nonpoor, and 7 that the poor are very much *more* materialistic; a scale point of 4 indicates that the two are equally materialistic.

Before we test a random group of middle-class adults, we set up two alternate hypotheses. The first is that in the population as a whole the mean scale point that will be chosen is 4—the hypothesis that the poor and nonpoor are believed to be equal. (From a statistical point of view we could have selected any value from 1 through 7 for our hypothesis, but for other reasons we have chosen 4.) This hypothesis, which specifies an exact value for μ, is known as the *null hypothesis* or H_0. The alternative hypothesis, H_1, is that μ is some value other than the one specified in the null hypothesis.[1]

The H_1 hypothesis can be either bidirectional or unidirectional. That is, we may hypothesize merely that the true value of μ is some value greater or smaller than that identified in the null hypothesis, or we may specify the *direction* in which we expect the difference; for example, μ is some value *less*, but not greater, than the one specified in H_0. The importance of the distinction between a unidirectional and a bidirectional alternative hypothesis

---

[1] The alternative hypothesis could be a specific value, let us say in our example a value of 2.00. However, it is more common to set up a more general H_1, specifying that μ may be *any* value that differs from H_0 in a given direction (unidirectional hypothesis) or either direction (bidirectional). For example, if H_0 states that $\mu = 64$, a unidirectional H_1 might be that μ is any value greater than 64; the bidirectional H_1 would state that μ is some value greater *or* smaller than 64.

will become clear later when we discuss the procedures used to evaluate H_0. We will see that we use a so-called *two-tailed* test when H_1 is bidirectional but may choose to use a *one-tailed* test when H_1 is unidirectional.

Let us suppose that if the poor and nonpoor are *not* believed to be equally materialistic by middle-class adults (that is, if H_0 that $\mu = 4.00$ is not true), we have no sound guess as to which group would be judged to be more materialistic. Our H_1 would therefore be bidirectional. Our two hypotheses are therefore:

H_0: $\mu = 4.00$

H_1: $\mu \neq 4.00$

We test 100 individuals and find that the mean scale point chosen is 3.91 and that the s of the sample is .70. We now proceed to evaluate the null hypothesis that $\mu = 4.00$.

What we want to do first is to determine what the sampling distribution of $\overline{X}$'s would look like if $\mu = 4.00$ and we had tested not one but a very large number of random samples of 100 cases each. We know that the mean would be 4.00, and we now must inquire about the shape of the sampling distribution and the SD of the distribution, $\sigma_{\overline{X}}$. We have no way of calculating $\sigma_{\overline{X}}$, but we may obtain an *estimate* of the standard error of the mean (which we have agreed to symbolize as $s_{\overline{X}}$) by using the sample s value given above:

$$s_{\overline{X}} = \frac{s}{\sqrt{N}} = \frac{.70}{100} = .07$$

The SD of the sampling distribution, we thus estimate, is .07. With a sample as large as the one we have obtained, the *shape* of our distribution is close enough to normal to permit us to assume the normal distribution. Our hypothetical sampling distribution of $\overline{X}$'s under H_0, then, is normal in shape, has a μ of 4.00 and an (estimated) SD of .07. We have shown this hypothetical distribution in Figure 8.2. We have also shown where our obtained $\overline{X}$ of 3.91 falls in this sampling distribution.

Our next task is to find the z score of our obtained $\overline{X}$ value by dividing the quantity $(\overline{X} - \mu)$ by SD. However, since we can only estimate SD, we can only estimate z. In this chapter and the next we will call this estimated z score a t score. The formula for this estimated z or t score is:

$$t = \frac{\overline{X} - \mu}{s_{\overline{X}}} \tag{8.5}$$

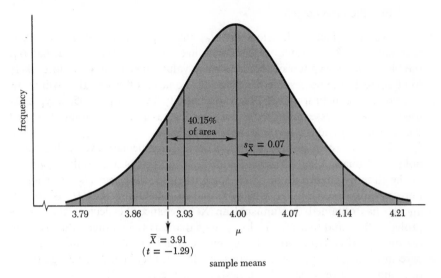

Figure 8.2 Visual test of the hypothesis that $\mu = 4.00$ when $\overline{X} = 3.91$ and $s_{\overline{X}} = .07$

This formula, of course, is the same we used earlier for z, except that we have substituted for the unknown standard error, $\sigma_{\overline{X}}$, our estimate of its value, $s_{\overline{X}}$. Applying this t formula to our data, we find:

$$t = \frac{\overline{X} - \mu}{s_{\overline{X}}} = \frac{3.91 - 4.00}{.07} = \frac{-.09}{.07} = -1.29$$

Our obtained $\overline{X}$, then, falls 1.29 $s_{\overline{X}}$ units below μ. Since with a sample as large as ours the sampling distribution can be assumed to be normal, we can treat this value as if it were a z score and determine from Table B the percentage of the area lying between μ and -1.29. Table B indicates that 40.15 percent of the area under the curve lies between μ and -1.29 and that therefore 9.85 percent $(50 - 40.15)$ of the area lies *beyond* -1.29. Since our alternative hypothesis, H_1, is bidirectional, we must be concerned with the other end of the distribution as well. We therefore state that if the null hypothesis, H_0, were correct, the probability of obtaining by chance a sample $\overline{X}$ that deviated from μ *this much or more* in either direction would be .197 $(.0985 \times 2 = .197)$. In other words, if we were to repeat the study over and over and over again, we would expect to get sample $\overline{X}$'s that deviated this much or more from $\mu = 4.00$ in either direction approximately 20 percent of the time.

Significance Levels

We now face a decision: is H_0, the hypothesis that $\mu = 4.00$, a reasonable one? Or should we conclude that it is unreasonable and that H_1, the bidirectional hypothesis that μ is some value other than 4, is more likely to be correct? We need some decision rule, some cutoff point, that will allow us to reach a conclusion. One of two probability levels, .01 or .05, is typically used in scientific research for this purpose. These are more commonly called the .01 (or 1%) and .05 (or 5%) *significance levels.*

If the investigator decides to use the 1% significance level as his cutoff point in evaluating H_0, he states that if $\overline{X}$ would be obtained only 1% *or less* of the time by chance alone if H_0 were true, he will *reject H_0* as being an unreasonable one. In other words, instead of concluding that H_0 is correct and he has obtained this unusual an $\overline{X}$ by chance, he decides that H_0 is probably false and that H_1, the hypothesis that μ is some other value, is more reasonable. If, on the other hand, his $\overline{X}$ could be obtained *more* than 1 percent of the time if H_0 were correct, he does not reject H_0, deciding instead to retain it as a reasonable hypothesis about the value of μ.

The 5% significance level sets a less stringent criterion than the 1% level for rejecting the null hypothesis about μ. If an investigator uses the 5% level, he will *reject H_0* if a difference as great as or greater than the difference between his obtained $\overline{X}$ and μ would occur by chance only 5 percent of the time or less if H_0 were true. Conversely, if an $\overline{X}$ this extreme would occur *more* than 5 percent of the time by chance, he will retain H_0 as a reasonable hypothesis.

In our study of beliefs about poverty, suppose that we had decided beforehand to use the 5% significance level in reaching a decision about the hypothesis that $\mu = 4.00$. (The significance level we set as our cutoff point is known as the *alpha* (α) *level.* Thus, when we are employing the 5% or .05 significance level, $\alpha = .05$ or equivalently, $\alpha = 5\%$.) We therefore will reject as unlikely any null hypothesis that would result in a sample $\overline{X}$ as different from μ as the one we have obtained 5 percent of the time or less by chance alone. We remind you that we have specified that H_1 is bidirectional; that is, we are concerned about deviations that occur 5 percent or less of the time in *either* direction from μ. In a normal distribution, 5 percent of the cases are expected to fall at or beyond ± 1.96 $s_{\overline{X}}$ units from μ—$2^{1}/_{2}$ percent at either extreme. When the α level is .05 and H_1 is a bidirectional hypothesis, we will reject any null hypothesis about μ that would make our sample $\overline{X}$ lie further away from μ than ± 1.96 $s_{\overline{X}}$ units. We reject it because there are only 5 chances in 100 that we would get an $\overline{X}$ that deviated this much or more if H_0 were correct. Conversely, we would retain, at the 5% significance level, any hypothesis about μ that would make it deviate less than ± 1.96 $s_{\overline{X}}$ units. We saw above that our sample $\overline{X}$ fell only 1.29 $s_{\overline{X}}$ units

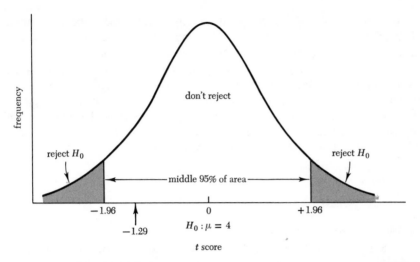

Figure 8.3 Illustration of a two-tailed significance test for μ at 5 percent level with normal curve values

from μ, an event that would occur more than 5% of the time if H_0 were true. We therefore *do not reject*, at the 5% significance level, the null hypothesis that $\mu = 4.00$; we accept H_0 as a reasonable one.

This procedure is illustrated in Figure 8.3, which shows a t distribution based on large samples and which we therefore have assumed to be normal. In this figure a center portion of the curve, bounded by $+1.96$ and -1.96 $s_{\bar{X}}$ units from μ, is marked off. If the value of any $\bar{X}$, expressed in $s_{\bar{X}}$ units (t score), falls within this area, we do not reject H_0; we accept H_0 as reasonable. If, on the other hand, the t score for $\bar{X}$ falls in either tail, we reject H_0. We saw above that our sample $\bar{X}$ fell only -1.29 $s_{\bar{X}}$ units from μ. We see in Figure 8.3 that this result falls within the "don't reject" area. We therefore retain, at the 5% significance level, the null hypothesis that $\mu = 4.00$; we conclude that H_0 is a reasonable one.

We could, of course, prepare a figure similar to Figure 8.3 for evaluating a bidirectional H_1 using the 1% alpha level. The extreme one-half percent of the area at either end of a normal curve falls at or beyond ± 2.58 SD units. The two tails of the distribution at or beyond ± 2.58 thus define the area of rejection.

When H_1 is bidirectional and we consider both extremes of the z or t distribution in evaluating H_0, we are using a *two-tailed* test. When we have good reason to believe before collecting the sample data that any deviation of $\bar{X}$ from the null hypothesis will be in a specific direction, we may elect to set up an H_1 that is unidirectional and to employ a *one-tailed* test, one that considers the probability of $\overline{X}$'s falling at only one extreme of the distribution. For example, suppose that in investigating beliefs about poverty we also ask our sample of 100 to compare the ability of the poor and the nonpoor to manage their money effectively. Again, our subjects use a 7-point scale, with 1 indicating that the poor are much less able to manage their money, 4 that the poor and nonpoor are equal, and so on. We also choose again to test the null hypothesis that $\mu = 4$. We suspect, however, that the poor will be rated as *less* able to manage money than the nonpoor; that is, we believe $\overline{X}$ will be less than 4. Our hypotheses are therefore:

$$H_0: \quad \mu = 4.00$$

$$H_1: \quad \mu < 4.00$$

(where $<$ means less than).

Suppose that the $\overline{X}$ turned out to be 3.75 and s to be .81. (The $\overline{X}$, you will notice, is in the direction specified by H_1.) Suppose also that we had decided to use $\alpha = .01$ instead of .05 as in the example above. The extreme 1 percent of the cases falling at *one* end of a normal distribution, we find in Table B, lie 2.33 $s_{\overline{X}}$ units or more from μ. We calculate the t for our $\overline{X}$ of 3.75 as follows:

$$t = \frac{\overline{X} - \mu}{s_{\overline{X}}} = \frac{3.75 - 4.00}{.081} = -3.09$$

where

$$s_{\overline{X}} = \frac{s}{\sqrt{N}} = \frac{.81}{\sqrt{100}} = .081$$

The t for our $\overline{X}$ falls far beyond -2.33. If H_0 were true, we would obtain an $\overline{X}$ this different from μ less than 1 percent of the time. We therefore *reject* the null hypothesis that $\mu = 4.00$ and accept as more reasonable the unidirectional alternate hypothesis that μ is less than 4.[2]

Statistical vs. Experimental Hypotheses

The example above may have puzzled you. If we had good reason to expect that middle-class individuals believe that the poor manage their money less well than the nonpoor, why did we test the particular null

---

[2] What if the difference between $\overline{X}$ and the μ specified in H_0 turns out to be in the *opposite* direction as specified in a unidirectional H_1? Obviously, H_1 can be rejected without further ado. Unfortunately, it is not legitimate to test H_0 by changing H_1 to a bidirectional form after the fact and performing a two-tailed significance test or to change the direction of H_1 to correspond to the results.

hypothesis that we did—that they are judged, by middle-class individuals as a group, to be *equally* capable? That is, why did we test the hypothesis that $\mu = 4$ rather than a value less than 4? The answer is that we suspected that μ has *some* value less than 4 but had no basis for predicting how much less. The null hypothesis, however, must be assigned an exact value. We therefore chose to select 4.00 (no difference between poor and nonpoor) as the null hypothesis to test, with the hope that we could *reject* this hypothesis and accept, instead, the more general H_1 hypothesis that μ is *some* value less than 4.00.

The null hypothesis that we test, then, may sometimes be picked only for statistical reasons. Our actual *experimental* hypothesis—our guess about the true state of affairs that we conducted the study to evaluate—may not always correspond to the null hypothesis. Of course, there are instances in which our experimental hypothesis specifies an exact value for μ, rather than a range of values (such as less than 4). In these instances the statistical null hypothesis that we test is the same as our experimental hypothesis.

One-Tailed vs. Two-Tailed Tests. When the null hypothesis that an investigator tests statistically does not coincide with his experimental hypothesis, he typically has predicted beforehand the direction in which he expects $\overline{X}$ will differ from the μ specified in H_0. There is some controversy among statisticians as to whether a unidirectional H_1 should be set up and H_0 evaluated by means of a one-tailed test when the investigator has predicted the outcome or whether all tests of H_0 should be two-tailed. We will not attempt here to discuss the reasons for the controversy. We will say only that one-tailed tests are sometimes used when the investigator has good reason to set up a unidirectional H_1. Despite the fact that investigators usually have unidirectional experimental hypotheses, bidirectional *statistical* hypotheses are much more common. It typically is assumed that H_1 is bidirectional and that a two-tailed significance test has been performed, unless a unidirectional H_1 and a one-tailed test are specifically mentioned. (We will follow this practice in all subsequent discussions.) In reporting the results of our statistical analyses, then, we should always specifically mention the use of a one-tailed probability figure if we have employed this procedure.

Unidirectional H_1's could be discussed in connection with other kinds of significance tests introduced in later chapters. We have chosen not to do so, however, both because no new principles are involved and because they are used less frequently than bidirectional H_1's.

Two Types of Error

Let us repeat once more what we have done when we reject a null hypothesis about μ at a given α level. For example, we reject a hypothesis at the 5% level. We say that if H_0 is correct, a difference between $\overline{X}$ and the hypothesized μ as large as or larger than the one we have obtained would occur by chance 5 percent of the time or less. Rather than concluding that by chance our sample happened to deviate this much from μ, we decide instead that the null hypothesis we tested is faulty. But we could be wrong. In fact, we know that the probability of being wrong is .05. If H_0 were *true* and we repeated our study over and over again, we would get $\overline{X}$'s that differed this much or more from μ 5 times out of 100. Thus, in rejecting the null hypothesis on a given occasion we run the risk of labeling as false a hypothesis that is in fact true. Rejecting a true hypothesis is known as a *Type I* error. The Type I error also is identified as an α error.

How might we reduce the likelihood of making a Type I error? The answer is simple enough: we set a more stringent α level for rejecting H_0. If we used $\alpha = .01$, we would demand that a deviation between $\overline{X}$ and μ occur by chance no more than 1 time in 100, thus cutting down the probability of rejecting a true hypothesis to .01. We would be even surer of avoiding a Type I error by using the .001 α level (only 1 chance in 1,000 of obtaining the results by chance), and so on. However, while we reduce the risk of a Type I error by lowering our α level, we simultaneously *increase* the likelihood of making what is called a Type II error. In a *Type II error* (also called a *beta* or β error), we retain (fail to reject) a hypothesis that is actually *false*. We lower the risk of a Type II in exactly the *opposite* way as we minimized a Type I error: by *raising* the α level. If we use the 5% level of significance, for example, we will reduce Type II error—that is, fail to reject false hypotheses less easily than if we used $\alpha = 1$ percent—and reduce this error even more if we raise α to 10 or 20 percent.

Although changing our α level has opposite effects on the likelihood of Type I and II errors, the probabilities of the two errors are not mathematically the exact complement of each other. That is, if α is 5%, the probability of a Type I error is .05 when we reject a hypothesis but is *higher* than .05 for a Type II error when we accept a hypothesis. You may begin to appreciate the reason for the greater frequency of Type II errors when you consider the following. Suppose we postulate μ to be 52 and find $\overline{X}$ to be 53. After testing this H_0, we conclude at a given α level that it is tenable. But we could have set H_0 as $\mu = 53$ or $\mu = 54$, among other values, and we would have concluded that these hypotheses are also tenable. In short, a whole *range* of null hypotheses would be acceptable at a given α level. When we retain a null hypothesis, all we are saying is that it is *one* of a number of acceptable hypotheses. (For this reason, investigators prefer not to speak of "accepting" the null hypothesis; they are more likely to be cautious and say that they have *failed to reject* the null hypothesis.)

In part because Type II errors are more common than Type I, we have a special concern with the ability of our statistical test to detect false null hypotheses. This ability is known as the *power* of the test, a topic that we discuss in more detail in the next chapter.

Having performed a study, we have no way of knowing whether we have committed a Type I or a Type II error, depending on whether we have accepted or rejected H_0. Nor is there any way of adjusting the α to reduce the probability of one type of error without increasing the probability of the other. The α that we set depends on the relative costliness of the two types of errors in our particular situation. We will postpone further discussion of this topic until the next chapter.

CONFIDENCE INTERVALS

In working with research data we often do not have hypotheses about specific values of μ that we wish to test. For example, in an earlier chapter we mentioned a scale designed to measure attitudes toward women's sex roles. Suppose we were interested in the attitudes of male chemistry professors and were able to persuade a random sample of 80 professors around the country to take the scale. We have no prior hypothesis about μ that we wish to test; instead, we are interested in establishing the *range* of values within which we may reasonably assert that μ lies. The assertion that μ falls somewhere within a specific range of values has a certain probability of being correct; we make it with a certain degree of confidence. When we establish such a range, it is commonly spoken of as a *confidence interval.* In research reports two confidence intervals are used most frequently: the *99% confidence interval* and the *95% confidence interval.*

Ninety-Nine % Confidence Interval

You will remember that if we were testing a hypothesis about μ using $\alpha = .01$ (and a bidirectional H_1), we would reject any H_0 that would make an absolute difference between μ and $\overline{X}$ as large as or larger than the one we have obtained occur 1 percent of the time or less by chance alone (that is, a positive difference between μ and $\overline{X}$ of the size we have obtained or larger would occur by chance one-half percent of the time or less and a negative difference of this size or larger one-half percent of the time or less). Conversely, we would retain as reasonable *any* H_0 that would make our $\overline{X}$ occur by chance more than 1 percent of the time. As we indicated in an earlier discussion, there is a whole range of reasonable or acceptable hypotheses at a given α level. If $\alpha = 1$ percent, the range of acceptable hypotheses for μ establishes what is known as the *99% confidence interval* for μ. That is, we assert that μ falls someplace within this range of values, with a probability of .99 (99 chances out of 100) that the statement is correct. (The probability that μ does *not* lie in this range is thus $1 - .99$ or $.01$.)

Let us now demonstrate how to go about establishing the 99% confidence interval. Earlier we mentioned giving a random sample of 80 professors an attitudes-toward-women scale. We find that $\overline{X}$ is 63, s is 15, and $s_{\overline{X}}$ is 1.68. We now would like to establish the 99% confidence interval for μ of the population.

If we were testing a hypothesis about μ, we would convert $\overline{X}$ into a t score and, assuming $\alpha = .01$, reject the hypothesis if t were greater than ± 2.58. (We use the figure 2.58 because we assume a normal sampling distribution with an N of this size; in a normal distribution, 99 percent of the $\overline{X}$'s fall within the range established by ± 2.58 SD units from μ and 1 percent beyond it.) How small could our hypothesis about μ be and still be accepted? We could answer this question by setting t at the value associated with $\alpha = .01$ (in this case, 2.58) and then entering our obtained $\overline{X}$ and $s_{\overline{X}}$ into the right-hand side of the t equation. Finally, we solve for μ:

$$t_{.01} = \frac{\overline{X} - \mu}{s_{\overline{X}}}$$

and by rearrangement:

$$\mu = \overline{X} - t_{.01}(s_{\overline{X}})$$

where $t_{.01} = t$ associated with $\alpha = .01$. Thus, for our example:

$$2.58 = \frac{63 - \mu}{1.68}$$

and

$$\mu = 63 - 2.58(1.68) = 63 - 4.33$$

The smallest μ that would be acceptable, with $\alpha = .01$, is $63 - 4.33$. We can find the largest acceptable μ by setting t at -2.58 and again solving for μ:

$$\mu = \overline{X} - t_{.01}(s_{\overline{X}}) = 63 - (-2.58)(1.68) = 63 + 4.33$$

These two values, $63 - 4.33$ (or 58.67) and $63 + 4.33$ (or 67.33), establish the limits of the 99% confidence interval. Since any hypothesis about μ between the values of 58.67 and 67.33 would be acceptable with $\alpha = .01$, we assert, with a .99 probability of being correct, that μ falls within this range of values.

If we generalize from this series of steps, we see that the 99% confidence interval for any set of data is given by:

$$\overline{X} \pm t_{.01}(s_{\overline{X}}) \tag{8.6}$$

When samples are large, so that we may assume that the sampling distribution of $\overline{X}$'s is normal, t is 2.58, the value in the normal curve associated with $\alpha = .01$.

Ninety-Five % Confidence Interval

The 95% confidence interval establishes the range of acceptable hypotheses for μ with $\alpha = .05$. When we set up the 95% interval, we assert that the probability is .95 that this range of values includes μ. We determine the 95% confidence interval exactly as we did the 99% interval, except that we now use the value of t associated with $\alpha = .05$. The formula for the 95% confidence interval is:

$$\overline{X} \pm t_{.05}(s_{\overline{X}}) \tag{8.7}$$

In a normal curve $\alpha = .05$ is associated with a z of ± 1.96. In our example of attitude scores the formula for the 95% confidence interval of μ is therefore:

$$\overline{X} \pm 1.96(s_{\overline{X}})$$

which we determine to be:

$$63 \pm 1.96(1.68) = 63 \pm 3.29$$

The 95% confidence interval for μ therefore is $59.71 - 66.29$. This range of values, you will notice, is narrower than the one we obtained for the 99% interval.

In using the 95% confidence interval to specify a range of values within which the mean of a population may lie we are less sure of including the true μ than with the 99% interval, but for the sake of greater specificity many research workers are willing to take the chance that μ does not fall beyond the limits set up by the 95% interval. The 95% confidence level, however, is usually considered the minimum probability value that should be used in setting up a range of values for μ. We could, for example, set up the 90% confidence interval. But then there would be 10 chances in 100, or 1 in 10, that μ lies outside this interval, and this probability of error is generally agreed to be too high. One could set up confidence intervals that fall between the 99 and 95% intervals, such as a 97% confidence interval. As a matter of custom, however, the 99 and 95% confidence intervals are commonly used as reference values. Similar remarks can be made about the use of the 5% and 1% significance levels in testing hypotheses about μ. Under ordinary circumstances the maximum α (and thus the minimal significance level) that should be used is 5 percent. If we go much beyond this level—for example, to the 10% level—we make the risk unacceptably high that we will commit a Type I error; that is, we will too frequently reject hypotheses about μ that are true.

The use of significance levels (which allow us to reject or not reject hypotheses with a stated probability) simplifies our writing of research reports. We merely state which one of the two α levels we are using and reject any hypothesis that exceeds this level. It is important to note once more that we must always state whether we are using the 5% or 1% α level in deciding to reject or not reject a hypothesis, so that there will be no misunderstanding. When we wish to establish a confidence interval for μ rather than to test a specific hypothesis about its value, we must also be sure to state whether it is the 99% or 95% interval we have determined.

STANDARD ERROR OF THE MEAN
AND SAMPLE SIZE

Earlier, when we described the central-limit theorem, we stated that as sample size increases, sample $\overline{X}$'s come closer and closer to μ in value. The theorem also states, more specifically, that the SD of the sampling distribution of $\overline{X}$'s is given by $\sigma/\sqrt{N}$. We will now explore further the implications of these points. Suppose that we measure the heights of women and have a very large number of samples of two cases each. The mean of the sampling distribution takes the same value as μ, but the individual sample $\overline{X}$'s are highly variable. It would not be too outlandish, for example, to obtain a sample of two short women whose mean height was little more than five feet or a sample of two tall women whose mean was close to six feet. But what if the samples each consist of 500 women? It would be *extremely* rare for sample $\overline{X}$'s based on 500 cases to be this different from μ. Almost without exception, the $\overline{X}$'s of samples this large would all cluster very closely around the mean height of the population as a whole.

These statements imply that the size of the standard deviation of the sampling distribution (the standard error of the mean or $\sigma_{\overline{X}}$) is inversely related to N. We can see that this is true by inspecting the formula for $\sigma_{\overline{X}}$: $\sigma/\sqrt{N}$. If the mean of a population were 120, for example, and the population SD were 10, then with samples of two cases each, $\sigma_{\overline{X}}$ would equal $10/\sqrt{2}$ or 7.07. But with N = 500, $\sigma_{\overline{X}}$ would equal $10/\sqrt{500}$ or only .45. As this example illustrates, as N gets larger and larger and approaches the size of the population, $\sigma_{\overline{X}}$ approaches zero.

Sample Size and Type II Error. This relationship between $\sigma_{\overline{X}}$ and N has implications for testing hypotheses about μ or setting up confidence intervals. When testing hypotheses about μ at a given α level, we are more likely to detect null hypotheses that are false, thus reducing the incidence of Type II errors, when the sample size is large than when it is small. In the illustration above, for example, in which $\mu = 120$, we might hypothesize, incorrectly, that $\mu = 122$. Testing a sample of two, we obtain $\overline{X}$ of 119 and s of 10. Thus, $s_{\overline{X}}$ would be 7.07 and t would be 3/7.07 or .424; we would fail to reject the false hypothesis that $\mu = 122$. With N $= 500$, on the other hand (and continuing to assume $\overline{X} = 119$ and $s = 10$), $s_{\overline{X}} = .45$. Therefore, t would equal 3/.45 or 6.67. With N $= 500$, we would *reject* the false hypothesis about μ far beyond the 1% level of significance. Large samples are also advantageous in establishing the 95% and 99% confidence intervals by allowing us to be more precise. When N is large, we identify a far narrower range of values within which we state that μ falls at a given level of confidence than when N is small.

CONFIDENCE INTERVALS AND SIGNIFICANCE LEVELS FOR SMALL SAMPLES

The central-limit theorem also states that as samples increase in size, the shape of the sampling distribution of $\overline{X}$'s approaches the normal curve. When N is large, we have said, the sampling distribution of t so closely approximates the normal curve that for practical purposes we may treat it as normal. In our examples, all of which involved large samples, we therefore used the z-score values found in the normal curve table in setting up the 99% and 95% confidence intervals about μ or in testing hypotheses using the 1% and 5% significance levels.

As samples become smaller and smaller, however, the t distribution remains symmetrical but increasingly departs from normality. When we have measured only a small sample—say fifty cases or less—and wish to set up a confidence interval or test a hypothesis about μ, it would not be appropriate to use the normal curve values in Table B to establish the significance levels of t's or to specify the limits of confidence intervals. Fortunately, the exact shape of the distribution of the t statistic is known, being indirectly related to sample size. Instead of normal curve values we use the values for the t statistic found in a special table (Table C at the end of the book). We will discuss this table and its use in testing hypotheses about μ in the next chapter.[3]

---

[3] For the reader's future reference we will anticipate a bit the discussion of Table C (t table) in Chapter 9. We enter Table C with $df = N - 1$ (number of cases in the sample minus one) and use the value associated with our df that is listed in the 5% column (two-tailed) in establishing the 95% confidence interval or in the one-tailed or two-tailed 5% column (depending on H₁) for testing hypotheses at the 5% significance level. We proceed similarly in determining the 1% significance level (two-tailed test) or 99% confidence interval. For example, if N is 25, then $df = 24$, and, after consulting Table C, we set up the 95% confidence as $\overline{X} \pm 2.064 s_{\overline{X}}$.

For the moment simply take note that the use of normal curve values, such as 1.96 and 2.58 in establishing the limits of the 95% and 99% confidence intervals or the two-tailed 5% and 1% significance levels, is appropriate for large samples but not for small ones.

STANDARD ERROR OF
A PROPORTION

Formulas are available to obtain standard errors for several statistics in addition to the mean, which permit us to test hypotheses about their population values. However, only one other standard error is used with sufficient frequency to warrant discussion here. This is the standard error of a proportion, symbolized as σ_{prop}.

Suppose we are investigating the biases that may influence choice behavior. It has been shown, for example, that in predicting heads or tails in the toss of a coin, people tend to pick heads; in guessing what number from 1 to 10 someone is thinking of, they tend to pick an even number; in a choice between right and left, they tend to choose right, and so on. We give a random sample of 50 individuals a novel series of tests, one of which involves asking them to guess whether the nonsense statement, "The sagem is bep," is true or false. If there is no bias, we expect that 50 percent of the subjects would choose true and 50 percent false. The null hypothesis (H_0) that we decide to test, then, is that the percent choosing true is 50. We find that the actual numbers are 32 choosing true and 18 choosing false—percentages of 64 and 36. What should we conclude?

By this time we are wise enough not to jump to the conclusion that H_0 is false and that there is a bias for choosing true simply because more than half of the sample chose this alternative. We recognize that there are sampling errors involved and that the question we really need to ask ourselves is, "If there is no bias—that is, if the population proportion choosing right is .50—how likely is it that we would obtain a proportion as high as .64 with 50 cases?"

You probably recognize that we are discussing a binomial sampling distribution in which we determine, in each sample, the proportion of individuals choosing true (p) and the proportion of individuals choosing false (q), where $p + q = 1.00$. In Chapter 7 we saw that when $p = .5$, the normal distribution provides a good approximation of the binomial distribution with moderate N's or greater (say 10 or more). It would be somewhat more precise to utilize the binomial distribution to answer our question, but, because expansion of the binomial is so laborious, it is far more convenient to use the normal distribution.

If we drew a very large number of random samples of the same size from a population in which $p = .5$ and for each obtained a proportion, we may thus assume (if N is large enough) that these sample proportions would be distributed normally. If we calculated SD of this distribution, we could make statements of probability about where any sample would fall. (If these statements have a familiar ring it is because we used the same language in discussing the sampling distribution of $\overline{X}$.) Thus, we need to know SD of the distribution of sample proportions—that is, the standard error of the proportion, σ_{prop}.

We recall from Chapter 7 that in a binomial sampling distribution in which we specify the frequency of series of events out of a total of N occasions the SD (the standard error of the distribution of frequencies or σ_f) is given by $\sqrt{Npq}$. However, we are looking for the standard error of a distribution of sample *proportions*. Since proportion is determined by dividing frequency by N, σ_{prop} may be obtained by dividing the expression for σ_f by N: $\sqrt{Npq}/N$. This may be reduced to $\sqrt{pq/N}$. Thus:

$$\sigma_{prop} = \sqrt{\frac{p_t q_t}{N}} \tag{8.8}$$

where: p_t = true population proportion

$q_t = 1 - p_t$

N = sample size

It is important to note that p in the formula for σ_{prop} is not the obtained *sample* proportion; it is the *true* proportion in a known or hypothesized population.

The question we ask in our bias experiment, we repeat, is, "If the true population proportion is .50, is it likely that our obtained proportion, .64, occurred by chance?" In answering the question, we first calculate σ_{prop}:

$$\sigma_{prop} = \sqrt{\frac{pq}{N}} = \sqrt{\frac{(.5)(.5)}{50}} = \sqrt{\frac{.25}{50}} = .071$$

We can visualize our hypothetical sampling distribution and determine where our obtained proportion falls. This has been done in Figure 8.4.

We can see by inspection that it would be quite unusual to get a sample proportion of .64 if the true proportion were .50. However, we need a more precise probability statement and must therefore convert our sample proportion (p_s) into a z score. (Since we know σ_{prop} of the distribution, note that

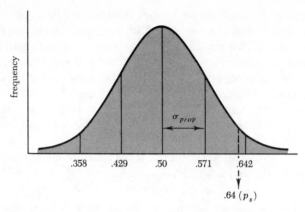

Figure 8.4 Visual test of the hypothesis that the true proportion is .50 when the obtained sample proportion is .64

we can calculate the actual z rather than an estimate of it.) Since the mean of the hypothetical sampling distribution is p_t, and the SD is σ_{prop}, we have:

$$z = \frac{p_s - p_t}{\sigma_{\text{prop}}} \quad\quad\quad (8.9)$$

$$= \frac{.64 - .50}{.071} = 1.97$$

where p_s = sample proportion and p_t = hypothetical true proportion.

If we had set α at .05 and our H_1 were bidirectional (that is, our alternate hypothesis to H_0 is $H_1: p_t \neq .50$), then we would reject any null hypothesis that would result in a z score, for our p_s, of ± 1.96 or greater. We use ± 1.96 as cutoff points, of course, because these are the values associated with the extreme 5 percent of the area in a normal distribution, $2^1/_2$ percent at each tail. Our z was calculated to be 1.97. We therefore reject the null hypothesis that $p_t = .50$; we conclude that the difference between p_s and p_t is significant at the .05 level. More concretely, we conclude that people are probably biased in favor of "true."

It might seem that for any problem in which we obtain a proportion we could test any hypothesis about the true proportion for a given N, using the same kinds of procedures as we have just illustrated. A word of caution is in order, however. The binomial distribution, we remind you, generally approaches the normal distribution as N grows larger, doing so most rapidly when p_t is .50. We therefore should not test hypotheses about p_t based on the normal curve assumption unless N is some reasonable size (10 at a minimum). As our hypothesis about p_t departs from .5, the sample N should be even larger than this minimum. If we test extreme hypotheses about true proportions (about .80 or greater or .20 or smaller), use of the normal curve may lead to serious errors of interpretation, when N's are not large, because the sampling distributions will be markedly skewed. Under this circumstance we can, however, fall back on the less convenient but more exact binomial distribution discussed in Chapter 7.

TERMS AND SYMBOLS TO REVIEW

Sampling distribution

Population

Sample

Representative

Sampling error of $\overline{X}$'s

Central-limit theorem

Standard error of mean $(\sigma_{\overline{X}})$

Estimated standard error of mean $(s_{\overline{X}})$

Null hypothesis (H_0)

Alternative hypothesis (H_1)

Unidirectional hypothesis

Bidirectional hypothesis

One-tailed test

Two-tailed test

Significance level (α level)

Statistical vs. experimental hypotheses

Type I (α) error

Type II (β) error

Confidence interval

t ratio

Standard error of proportion (σ_{prop})

Population proportion (p_t)

Sampling Distributions

Independent Samples from Two Populations

9

In Chapter 8 we learned that if we were to measure an infinite number of random samples of a given size and find $\overline{X}$ for each, the mean of the sampling distribution would be equal to the mean of the population, μ. We also learned that if the samples were relatively large, the shape of the sampling distribution would approach the normal curve. With this information and a way to estimate the SD of the sampling distribution (the standard error of the mean) we can test hypotheses about μ or set up confidence intervals about a range of values that may contain μ. In this chapter such reasoning will be extended to the sampling distribution of the *differences* between the $\overline{X}$'s of *pairs* of random samples drawn from two different populations. Knowing the properties of the sampling distribution of the difference between $\overline{X}$'s, we will be able to test hypotheses about the difference between population μ's ($\mu_1 - \mu_2$).

By testing such a hypothesis we can answer the question that psychological experiments involving two groups or conditions set out to answer. In a typical experiment one group of subjects is treated in one way (sample from first population) and another group in a different way (sample from second population). We perform the experiment to learn something about the relative effects of the two conditions on the subjects' reactions—that is, about the difference between the μ's of the two populations. We use the sample data we have obtained to test a hypothesis about this difference.

SAMPLING DISTRIBUTION OF THE DIFFERENCE BETWEEN MEANS

In discussing the sampling distribution of the difference between means, we must keep *two* populations in mind—not a single population as in Chapter 8. To illustrate, suppose we are interested in whether or not men and women differ in their memory for faces. We select two random samples of equal size, one composed of men and the other of women. After giving the subjects a memory test we find the mean performance of each group. We next determine the difference between $\overline{X}$'s, let us say the women's $\overline{X}$ ($\overline{X}_W$) minus the men's ($\overline{X}_M$). We could repeat this procedure very many times (technically, an infinite number), on each occasion finding the difference between the pair of the $\overline{X}$'s in the same order, $\overline{X}_W - \overline{X}_M$. Assuming that the populations are normally distributed or that the size of the samples is relatively large, we expect the shape of the resulting sampling distribution of differences between $\overline{X}$'s to approximate a normal distribution, just as in the case of the sampling distribution of $\overline{X}$'s from a single population. We also anticipate that the mean of the distributions of differences between pairs of sample $\overline{X}$'s will take the same value as the difference between the two population μ's (in our illustration, $\mu_W - \mu_M$, or more generally, $\mu_1 - \mu_2$). If we knew the standard deviation of the sampling distribution (which we will call the *standard error of the difference between $\overline{X}$'s* or $\sigma_{\overline{X}_1 - \overline{X}_2}$), we could reconstruct the sampling distribution and determine the probability of obtaining any given difference between $\overline{X}$'s of a given size or larger.

Standard Error of the Difference Between Means

The standard error of the difference, $\sigma_{\overline{X}_1 - \overline{X}_2}$, could be computed directly from the actual distribution of differences between $\overline{X}$'s. However, a constant relationship has been demonstrated between this standard error and the standard error of the mean ($\sigma_{\overline{X}}$) of each of the populations from which the samples are drawn; $\sigma_{\overline{X}}$ is in turn related to the σ of the population and the sample N, as we learned earlier. This relationship is shown in the following formula:

$$\sigma_{\overline{X}_1 - \overline{X}_2} = \sqrt{\sigma_{\overline{X}_1}{}^2 + \sigma_{\overline{X}_2}{}^2} \tag{9.1}$$

where

$$\sigma_{\overline{X}_1} = \frac{\sigma_1}{\sqrt{N_1}} \quad \text{and} \quad \sigma_{\overline{X}_2} = \frac{\sigma_2}{\sqrt{N_2}}$$

Using formula (9.1), we can determine $\sigma_{\overline{X}_1 - \overline{X}_2}$, the standard error of the difference between $\overline{X}$'s, if we know the σ of each population and the sample size, N.

Suppose that in our illustration about women's memory for faces vs. men's the population difference ($\mu_W - \mu_M$) is 3.00 and that, for a given sample N, $\sigma_{\overline{X}_1 - \overline{X}_2}$ is 1.5. The sampling distribution of differences between $\overline{X}$'s is illustrated in Figure 9.1. We now measure a single pair of random samples of N cases each and find the difference between $\overline{X}$'s ($\overline{X}_W - \overline{X}_M$) to be 1.6. How probable is it that we would obtain a difference between $\overline{X}$'s that deviates this much or more, in either direction, from the mean of the sampling distribution? We can see from Figure 9.1 that this ($\overline{X}_W - \overline{X}_M$) value lies less than one SD unit from the mean of the distribution; a deviation of this magnitude or larger would be a quite probable event. We can determine the probability more exactly by computing a z score for our obtained difference between $\overline{X}$'s and by then consulting the normal curve table. The general formula for determining z in sampling distributions of differences between means is:

$$z = \frac{(\overline{X}_1 - \overline{X}_2) - (\mu_1 - \mu_2)}{\sigma_{\overline{X}_1 - \overline{X}_2}} \tag{9.2}$$

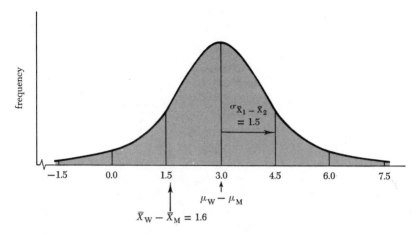

Figure 9.1   Hypothetical distribution of differences between sample $\overline{X}$'s for women's and men's memory for faces when the difference between population μ's ($\mu_W - \mu_M$) is 3.00 and the standard error is 1.5.

thus yielding for our example:

$$z = \frac{1.6 - 3.0}{1.5} = \frac{-1.}{1.5} = -.93$$

Looking at Table B, we find that 32.38 percent of the area in a normal curve falls between the mean and a z of .93 and 17.62 percent beyond this point; z scores exceeding $\pm.93$ therefore will occur approximately 35 percent of the time (17.62% × 2).

We rarely know μ and σ for a single population, let alone for two populations. Therefore, we cannot obtain either ($\mu_1 - \mu_2$) or $\sigma_{\overline{X}_1 - \overline{X}_2}$. However, if we had some way to estimate $\sigma_{\overline{X}_1 - \overline{X}_2}$, we would be able to test hypotheses about the difference between μ's, based on the data from a single pair of samples. Fortunately, we can arrive at such an estimate by using (in place of each $\sigma_{\overline{X}}$) estimated standard errors of the mean ($s_{\overline{X}}$), which we learned to calculate in Chapter 8.

Estimating the Standard Error of the Difference

The formula for estimating the standard error of the difference between $\overline{X}$'s parallels formula (9.1) given above for determining $\sigma_{\overline{X}_1 - \overline{X}_2}$. Thus:

$$s_{\bar{X}_1 - \bar{X}_2} = \sqrt{s_{\bar{X}_1}^2 + s_{\bar{X}_2}^2} \tag{9.3}$$

where

$$s_{\bar{X}_1}^2 = \frac{s_1^2}{N_1} \quad \text{and} \quad s_{\bar{X}_2}^2 = \frac{s_2^2}{N_2}$$

The calculation of $s_{\bar{X}_1 - \bar{X}_2}$ is illustrated in the section below.

TESTING A HYPOTHESIS ABOUT
$\mu_1 - \mu_2$: t RATIO

We perform an experiment with laboratory animals to determine whether a diet deficient in protein affects their general activity level. One group of 60 animals is fed a low-protein diet for six months; another group of 60 is given a normal amount of protein. Near the end of this interval members of both groups are tested for activity level. The $\bar{X}$ of the normal group turns out to be 97.51, and s is 13.7; for the deficient group $\bar{X}$ is 91.36 and s is 14.2.

Our purpose in testing these samples is to make an inference about *populations*—populations of animals given diets that are normal or deficient in protein. We decide to test the hypothesis that the μ's of the two populations are the *same*; in other words, our null hypothesis for this study is $\mu_1 - \mu_2 = 0$. Our alternate hypothesis is the bidirectional hypothesis that the difference is some value other than zero. Thus:

$$H_0: \quad \mu_1 - \mu_2 = 0$$

$$H_1: \quad \mu_1 - \mu_2 \neq 0$$

Which population is identified as 1 and which is 2 is purely arbitrary; in our example, 1 will refer to the normal diet condition and 2 to the deficient.

To test the hypothesis, we first must calculate the estimated standard error of the difference:

$$s_{\bar{X}_1 - \bar{X}_2} = \sqrt{s_{\bar{X}_1}^2 + \frac{s_{\bar{X}_2}^2}{N_2}} = \sqrt{\frac{s_1^2}{N_1} + \frac{s_2^2}{2}}$$

$$= \sqrt{\frac{(13.7)^2}{60} + \frac{(14.2)^2}{60}} = \sqrt{3.13 + 3.36}$$

$$= \sqrt{6.49} = 2.55$$

Next, we must obtain the estimated z score of our obtained difference between $\bar{X}$'s. We recall from Chapter 8 that the estimated z is identified as a t. The t formula for the difference between means is:

$$t = \frac{(\bar{X}_1 - \bar{X}_2) - (\mu_1 - \mu_2)}{s_{\bar{X}_1 - \bar{X}_2}} \tag{9.4}$$

In our example we proposed the null hypothesis that $\mu_1 - \mu_2 = 0$. When this particular H_0 is being tested, the expression $(\mu_1 - \mu_2)$ can be omitted from the numerator, so that the formula for t becomes:

$$t = \frac{\bar{X}_1 - \bar{X}_2}{s_{\bar{X}_1 - \bar{X}_2}} \tag{9.5}$$

where H_0: $\mu_1 - \mu_2 = 0$.

For reasons that will be discussed shortly, the null hypothesis almost always tested about $\mu_1 - \mu_2$ is that the difference is zero. That is, for statistical purposes the assumption is made that there are no differences between groups or conditions, so that $\mu_1 = \mu_2$. This assumption is so common that the expression "the null hypothesis" often is used synonymously with the specific hypothesis that $\mu_1 - \mu_2 = 0$; and the formula for t is frequently the one shown above in which the term $(\mu_1 - \mu_2)$ is omitted.

Application of formula (9.5) to the data from our groups of laboratory animals gives us:

$$t = \frac{97.51 - 91.36}{2.55} = \frac{6.15}{2.55} = 2.41$$

Significance Levels of t. As we described in Chapter 8 when discussing the testing of hypotheses about μ, we evaluate the null hypothesis about the difference between means by comparing our obtained t with the t associated with a predetermined α level. As before, α typically is either the 5% or the 1% level. When the obtained t equals or exceeds the t associated with our specified α level, we conclude that our t is *significant*. We *reject* the null hypothesis that we have set up about the difference between μ's and conclude that our alternate hypothesis is more reasonable.

When the sample N's are large, and the sampling distribution of t may be assumed to be normal (we recall from Chapter 8), the values associated with the 5% and 1% significance levels are those found in the normal curve table. We also remind you that when H_1 is bidirectional, these values are 1.96 and 2.58 for the 5% and 1% levels, respectively.

If α were set at the 5% level in our example, we would reject our null hypothesis that $\mu_1 - \mu_2 = 0$, since t was found to be 2.41. We would conclude that our obtained difference between $\bar{X}$'s was significant—that in the population as a whole, prolonged protein deficiency leads to lowered activity level.

Choice of the Null Hypothesis

If an investigator anticipates that the experimental conditions he investigates have no differential effects on his subjects' reactions, he assumes that the two population μ's are the same. Therefore, the null hypothesis he tests is $\mu_1 - \mu_2 = 0$. In most experiments, however, the investigator expects that the conditions he is studying *do* differentially affect his subjects; he does the experiment because he suspects that the population μ's are *not* equal. However, even when investigators have firm grounds for postulating not merely a difference between μ's but also the direction the difference may take, they rarely are willing to specify an exact value for this difference. But the logic of hypothesis testing, we noted previously, demands that H_0 be assigned a precise value. The solution is to adopt the statistical hypothesis that $\mu_1 - \mu_2 = 0$ in the expectation that it will be *rejected*.

THE t RATIO AND SMALL SAMPLES

Standard Error of Difference and Unequal N's

In our discussion of t thus far, we have dealt only with relatively large samples—samples that have 50 to 60 members or more. We make several modifications in our procedures when we wish to determine the significance of the difference between the means of small samples. First, we should use formula 9.3 for $s_{\bar{X}_1 - \bar{X}_2}$ *only if the N's for the two samples are equal*. To repeat this formula and the conditions under which it is used in comparing small samples:

$$s_{\bar{X}_1 - \bar{X}_2} = \sqrt{s_{\bar{X}_1}{}^2 + s_{\bar{X}_2}{}^2}$$

(small samples with equal N's)

In this formula, observe that the $s_{\bar{X}}$ of each sample, which in turn is based on the s (or $\tilde{\sigma}$) of each sample, contributes equally to our estimate of the standard error of the difference. When the sample N's are not equal, we must modify our formula so that the contribution of the data from each sample to our estimate of the standard error of the difference is weighted according to the sample N. This modification *must* be used for small samples with unequal N's but also may be used when N's are the same. Two versions of the modified formula are possible, depending on whether $\tilde{\sigma}$ or s has been computed for each sample:

$$\begin{aligned} s_{\bar{X}_1 - \bar{X}_2} &= \sqrt{\frac{N_1\tilde{\sigma}_1{}^2 + N_2\tilde{\sigma}_2{}^2}{(N_1 + N_2 - 2)}\left(\frac{1}{N_1} + \frac{1}{N_2}\right)} \\ &= \sqrt{\frac{(N_1 - 1)s_1{}^2 + (N_2 - 1)s_2{}^2}{(N_1 + N_2 - 2)}\left(\frac{1}{N_1} + \frac{1}{N_2}\right)} \end{aligned} \qquad (9.6)$$

(small samples with equal or unequal N's)

A raw-score formula also may be used:

$$s_{\bar{X}_1 - \bar{X}_2} = \sqrt{\frac{(\sum X_1^2 + \sum X_2^2) - (N_1\bar{X}_1^2 + N_2\bar{X}_2^2)}{(N_1 + N_2 - 2)} \left(\frac{1}{N_1} + \frac{1}{N_2}\right)} \quad (9.7)$$

Note that when sample N's *are* equal, these formulas produce exactly the same answer as does the simpler formula based on the addition of the two $s_{\bar{X}}^2$'s. Thus, as we have noted, the use of these modified formulas always is appropriate, whether the sample N's are the same or different.

Use of the *t* Table

A second difference in the procedures we employ with small samples has to do with interpreting the *t* that we have calculated. As we discussed earlier, the *t* values obtained from large samples approximate the normal distribution. Thus, in interpreting large-sample *t*'s, we use Table B, the normal curve table. However, the *t* values obtained from small samples are *not* normally distributed. Assuming that the populations from which the samples were drawn are normally distributed, the distribution of small sample *t*'s is, however, symmetrical and varies with what is known as the *degrees of freedom* associated with the samples. The number of *degrees of freedom* (*df*), we shall see shortly, is related to sample size. Thus, the shape of the sampling distribution of *t* depends indirectly on the number of cases in the samples.

The upshot of all this is that we need a special table when evaluating *t* ratios based on small samples. Such a table is represented as Table C at the end of the book. At the left of Table C we find a column labeled *df* (*degrees of freedom*). In the other columns are listed the *t* values for each *df* that are required at the 10%, 5%, 2%, and 1% significance levels for a two-tailed test and at the 5%, 2.5%, 1% and .5% significance levels for a one-tailed test.

The *df*, we have said, is based on N, and is obtained by either of two equivalent formulas:

$$df = N_1 - 1 + N_2 - 1 \quad \text{or} \quad df = N_1 + N_2 - 2 \quad (9.8)$$

If we draw two independent samples where N is 10 each, the *df* is 18 (10 + 10 − 2). If we have two samples, one of 7 and the other of 9 cases, the *df* is 14.

As *df*'s become smaller, the shape of the sampling distribution of *t*'s becomes increasingly flatter than the normal curve. This increasing flatness means that we must go further and further along the *t* scale on the baseline (in other words, the estimated *z* scale) to encompass the same area as in a normal curve. In a normal curve, the extreme 5 percent of the area $(2^1/_2$ percent at each end), for example, lies beyond ± 1.96 units. This t value will obtain (for all practical purposes) when df is very large, and we may use it to specify the 5% significance level. When df is only as large as 100, for example, we observe in Table C that the t required for the 5% significance level is 1.9840; only slightly larger than 1.96. As df becomes smaller, the critical value of t becomes larger; for $df = 30$, it is 2.042; for $df = 10$, it is 2.228; and so on.

Two Examples. We can now demonstrate how to go about testing the significance of the difference between $\overline{X}$'s of small samples. Suppose we test five thirsty rats in a maze and give them a drink of water after each trial. Another group of five is given a saccharine solution to drink after each trial. Our experimental question is whether learning the maze will be more rapid with one reward than with the other. The null hypothesis we test is that the rewards have equal effects on performance; the alternate hypothesis is that they do not. Therefore:

$$H_0: \quad \mu_1 - \mu_2 = 0$$

$$H_1: \quad \mu_1 - \mu_2 \neq 0$$

We have reported below the mean number of trials taken by members of each group to go through the maze without error and, for illustrative purposes, both the $\tilde{\sigma}^2$ and s^2 that we have calculated from the data of each sample. Since the two groups have the same N's, we use the simpler equal-N formula (9.3) for computing $s_{\overline{X}_1 - \overline{X}_2}$. To demonstrate, we will compute the $s_{\overline{X}}^2$'s from each sample (the basic components of $s_{\overline{X}_1 - \overline{X}_2}$) in two ways, first using the sample $\tilde{\sigma}$'s and then the sample s's.

Water	*Saccharine*
$N_1 = 5$	$N_2 = 5$
$\overline{X}_1 = 18.40$	$\overline{X}_2 = 15.60$
$\tilde{\sigma}_1^2 = 7.96$	$\tilde{\sigma}_2^2 = 6.64$
$s_1^2 = 9.95$	$s_1^2 = 8.30$
$s_{\overline{X}_1}^2 = \dfrac{\tilde{\sigma}_1^2}{N_1 - 1} = \dfrac{s_1^2}{N_1}$	$s_{\overline{X}_2}^2 = \dfrac{\tilde{\sigma}_2^2}{N_2 - 1} = \dfrac{s_2^2}{N_2}$
$= \dfrac{7.96}{5 - 1} = \dfrac{9.95}{5}$	$= \dfrac{6.64}{5 - 1} = \dfrac{8.30}{5}$
$= 1.99$	$= 1.66$

$$s_{\bar{X}_1 - \bar{X}_2} = \sqrt{s_{\bar{X}_1}^2 + s_{\bar{X}_2}^2} = \sqrt{1.99 + 1.66}$$

$$= \sqrt{3.65} = 1.91$$

$$t = \frac{\bar{X}_1 - \bar{X}_2}{s_{\bar{X}_1 - \bar{X}_2}} = \frac{18.40 - 15.60}{1.91} = \frac{2.80}{1.91}$$

$$= 1.47$$

$$df = N_1 + N_2 - 2 = 5 + 5 - 2 = 8$$

Having determined t and df, we now look at Table C. Assuming that we have chosen $\alpha = 5\%$, we determine that with 8 df we need a t of 2.31 to be significant at the 5% level. Since our t was only 1.47, we cannot assert with any confidence that the two rewards produced any difference in the speed of acquiring the maze. In short, we cannot reject the null hypothesis that $\mu_1 - \mu_2 = 0$.

Our second example comes from a study in which the perceptual-motor performance of individuals with a certain type of brain injury was tested. We wish to compare the mean performance of a group of twelve persons whose injury was unilateral—on only one side of the cortex—with the performance of a group of five whose injury was bilateral. You will notice that, since the groups do not have the same number of cases, we must use the formula appropriate for unequal N's when computing $s_{\bar{X}_1 - \bar{X}_2}$.

Unilateral	Bilateral
$N_1 = 12$	$N_2 = 5$
$\bar{X}_1 = 49.19$	$\bar{X}_2 = 38.43$
$\tilde{\sigma}_1^2 = 74.65$	$\tilde{\sigma}_2^2 = 49.28$

$$s_{\bar{X}_1 - \bar{X}_2} = \sqrt{\frac{12(74.65) + 5(49.28)}{12 + 5 - 2}\left(\frac{1}{12} + \frac{1}{5}\right)}$$

$$= \sqrt{\frac{895.80 + 246.40}{15}(.083 + .200)}$$

$$= \sqrt{(76.15)(.283)} = \sqrt{21.55} = 4.64$$

$$t = \frac{\bar{X}_1 - \bar{X}_2}{s_{\bar{X}_1 - \bar{X}_2}} = \frac{49.19 - 38.43}{4.64}$$

$$= \frac{10.76}{4.64} = 2.32$$

$$df = 12 + 5 - 2 = 15$$

We find in Table C that with 15 *df* a *t* of 2.13 is required for significance at the 5 percent level and 2.95 at the 1 percent level. If α were .05, we would conclude that the difference between the groups was significant; we would reject the null hypothesis that $\mu_1 - \mu_2 = 0$, concluding instead that bilateral brain injury results in poorer perceptual-motor performance than unilateral injury. If α were .01, on the other hand, we would not reject the hypothesis.

Testing Hypotheses About *μ* with Small Samples

We will now digress from the main topic of this chapter (testing the significance of the difference between $\overline{X}$'s) and go back to the procedures discussed in Chapter 8 for testing hypotheses about the *μ* of a single population. As mentioned there, we use the normal curve table in determining the values associated with the 5% and 1% significance levels only if samples are large. When samples are small, we must use the *t* distribution. In testing a hypothesis about *μ* based on the data of a small sample, we enter Table C with *df* = N − 1 (the number of cases in the sample minus one). Having found the value at the 5% or 1% significance level for our *df*, we compare it with our calculated *t* (obtained, you recall, by dividing $\overline{X} - \mu$ by $s_{\overline{X}}$). If our calculated *t* is smaller than the tabled value, we accept the hypothesis about *μ* at the given significance level, but if it is greater than the tabled value, we reject the hypothesis. We also use the (two-tailed) values found in Table C in establishing the 99% and 95% confidence intervals for *μ*, again entering the table with *df* = N − 1.

Assumptions Underlying *t*

We mentioned earlier, but did not emphasize, that the theoretically derived *t* distributions are based on the assumption that the characteristic being measured is normally distributed in the populations from which the samples were drawn. Use of the *t* ratio to test a hypothesis about the difference between population means also assumes that the population σ's are the same. It has been demonstrated, however, that *t* tests may be used without noticeably distorting our conclusions even when these assumptions are not met, as long as the σ's are not markedly different or the distributions do not depart radically from normal. A final assumption is that our two samples are both randomly drawn from their respective populations and are independent of each other. If one or more of these assumptions is seriously violated, one procedure that is open to us is to use an alternate type of statistical test that makes no assumptions about distribution shapes or σ's. Some of these *nonparametric* techniques, as they are called, are discussed in Chapter 15.

CHOICE OF SIGNIFICANCE LEVEL

Statisticians advise investigators to specify beforehand the α level they will use to decide whether or not to reject the null hypothesis they have designed their study to test. This procedure makes the decision-making process automatic and invulnerable to the investigator's biases or wishful thinking. The investigator does, however, have a choice of significance level, which raises a question about the factors that determine this choice. We begin our discussion of this problem by presenting two examples.

The research division of a drug company has discovered a new drug for the treatment of a specific medical disorder. It is not known whether the drug differs in its therapeutic effectiveness from the standard remedy, but it is known that the two drugs are otherwise comparable in cost, ease of administration, side effects, and so on. An experiment is therefore planned to compare the two drugs' therapeutic effectiveness. The null hypothesis the investigators decide to test is that the two drugs have equal effects ($\mu_1 = \mu_2$); the alternate hypothesis they set up is the unidirectional hypothesis that the new drug is *better*. If the new drug turns out to be significantly better than the old, the company will put it on the market. In considering which α level to use, the researchers ponder the possible kinds of decisions that could result. The H_0 is either true or false, and, after assessing H_0 statistically, they will either accept or reject it. Thus, there are four possibilities, two of them involving ways to come to a *correct* conclusion: the investigator can reject H_0 when it is in fact false and can accept H_0 when it is in fact true. Similarly, there are two ways to come to an *incorrect* conclusion: the investigator can erroneously reject H_0 when it is in fact true (Type I error) and can accept H_0 when it is in fact false (Type II error). These possibilities and the probabilities of each are shown in Table 9.1. Recall from Chapter 8 that the probability of making a Type I error at a given α level is equal to that α,

Table 9.1   The four possible outcomes in evaluating H_0 at a given α level and probabilities of each.

		Correctness of H_0	
		True	False
Conclusion	Accept H_0	Correct $1 - \alpha$	Error (Type II) β
	Reject H_0	Error (Type I) α	Correct $1 - \beta$

while the probability of a Type II error takes a value somewhat higher than α and is identified as β.

The researchers must weigh the hazards of the two types of errors. If the new drug is better than the old and they accept the false hypothesis that $\mu_1 - \mu_2 = 0$, the drug will not be marketed and those who suffer from the disorder will be deprived of a more potent therapeutic agent. On the other hand, if the new drug is no better than the old and they incorrectly *reject* H_0, the company will waste time and money marketing the new product. Marketing expense, however, is quite minimal, and they decide that a Type I error (concluding that the new drug is more effective when it isn't) would be less costly than a Type II error (concluding that the new drug is equally effective as the old when it is better). Therefore, they decide to use a relatively large value of α, 5 percent or perhaps even 10 percent, to minimize the probability of a Type II error.

Another example: a new method of teaching eighth-grade biology has been devised that would be quite expensive to implement, requiring new books, special laboratory equipment, and retraining of teachers. Before deciding whether or not to introduce the method into a large elementary school, the principal asks that an experiment be conducted in which the exam scores of a group of pupils taught by this method are compared with those of a group taught in the regular way. Because of the expense involved, an α level of 1 percent or even smaller might be used to assess H_0. Setting a more stringent criterion for rejecting the null hypothesis minimizes the probability of a Type I error—that is, lessens the probability that the H_0 will be incorrectly rejected when in fact the new method does not produce better performance than the old. (The principal also might insist that the new method produce results not only *better* than the old, but *substantially superior*. He therefore might ask that H_0 be set at some value other than 0.)

In both of these examples a practical decision is to be based on the conclusion drawn from the experimental data. The α level chosen for assessing the null hypothesis is adjusted up or down, depending on the investigator's judgment about the relative risks of a Type I or Type II error. In purely scientific research the acceptance or rejection of the null hypothesis usually does not have consequences as immediate and as obvious as those in our examples. Nonetheless, errors can be costly. Promising ideas can be abandoned or investigators' time and energy wasted in pursuit of false trails as a consequence of an incorrect decision about H_0. In general, however, scientists prefer to be conservative about rejecting hypotheses and consider a rule that would permit the occurrence of a Type I error more than 5 percent of the time to be unacceptable. Therefore, the highest α value conventionally used is .05. On the other hand, setting α lower than .01 results too often in failure to detect a false H_0 (Type II error) to be acceptable. One therefore finds the 1% and 5% levels used almost exclusively in research reports.

Since there usually is little pressure on anyone to take direct action based on an investigator's decision about the correctness of H_0, most investigators report their data as being significantly different from the null hypothesis if it can be rejected at the 5% level or less. If it is possible to reject H_0 at the 1% level, this is the significance level that is actually reported rather than 5%.

POWER OF A STATISTICAL TEST

In their concern not to make Type I errors, scientists have specified relatively stringent criteria for rejecting the null hypothesis, the 5% and 1% significance levels. They are also concerned about minimizing Type II errors and hence about the probability that a statistical test will detect false hypotheses. This probability, known as the *power* of the test, is determined by the following formula:

$$\text{power} = 1 - p(\text{Type II error}) \tag{9.9}$$

We have noted that the probability of a Type II error is symbolized by the Greek letter β (beta). Thus we may also write:

$$\text{power} = 1 - \beta$$

What the formula indicates, we repeat, is the probability that H_0 will be rejected when it is in fact false.

We will discuss power for the simple case in which we test H_0: $\mu_1 - \mu_2 = 0$ and know the values of μ and σ. We can illustrate the method of calculating power by the following example involving two populations. For the first population μ is 93 and for the second μ is 89; both populations have a σ of 8.

We test a random sample of 10 cases from each population and set up the null hypothesis that $\mu_1 - \mu_2 = 0$. (Note that the true difference is $93 - 89$ or 4 and H_0 is false.) For convenience, we will assume that our alternative hypothesis, H_1, is the unidirectional hypothesis that $\mu_1 > \mu_2$. The α level we decide to use in evaluating H_0 is .05.

Since we know the population σ, we are able to compute the z score for the difference between our sample $\bar{X}$'s (rather than having to obtain t, an estimate of z). The formula we use is:

$$z = \frac{(\bar{X}_1 - \bar{X}_2) - (\mu_1 - \mu_2)}{\sigma_{\bar{X}_1 - \bar{X}_2}}$$

Looking at the normal curve table, we see that 45 percent of the area lies between μ and 1.65 z and 5 percent falls beyond this value of z. Thus, with $\alpha = .05$, we will reject the null hypothesis that $\mu_1 - \mu_2 = 0$ if we obtain a z of 1.65 or greater.

We know that H_0 is false. The power of our test, which we have defined as the probability of detecting a false null hypothesis, is thus equal to the probability of obtaining a z of 1.64 or greater. We proceed to determine this probability by first determining $\sigma_{\overline{X}_1 - \overline{X}_2}$:

$$\sigma_{\overline{X}_1 - \overline{X}_2} = \sqrt{\sigma_{\overline{X}_1}{}^2 + \sigma_{\overline{X}_2}{}^2} = \sqrt{\frac{8^2}{10} + \frac{8^2}{10}}$$

$$= \sqrt{\frac{128}{10}} = 3.58$$

Next we find the "raw-score" value for a z score of 1.64—that is, the value of $(\overline{X}_1 - \overline{X}_2)$ associated with this z score:

$$1.64 = \frac{(\overline{X}_1 - \overline{X}_2) - 0}{3.58}$$

so that

$$(\overline{X}_1 - \overline{X}_2) = 1.64(3.58) = 5.87$$

We have now determined that if $(\overline{X}_1 - \overline{X}_2)$ is 5.87 or greater, we will reject H_0 and accept instead our H_1 that $\mu_1 > \mu_2$. This is illustrated in Figure 9.2.

Our next step is to determine the probability of getting a difference between $\overline{X}$'s of this magnitude or greater in the sampling distribution under H_1—that is, the *actual* distribution that would be obtained with our μ's of 93 and 89. We first compute z:

$$z = \frac{5.87 - (93 - 89)}{3.58} = \frac{5.87 - 4.00}{3.58} = .52$$

Looking at Table B, we find 19.85 percent of the area lies between μ and $z = .53$, and 30.15 percent beyond it. Thus, the probability of rejecting the false H_0 is .3015. This probability is equal to $1 - \beta$, *the power of the test*. We can also see that β, the probability of making a Type II error, is $1 - .3015$ or .6985. We add that a power of .3015 is usually considered too low a value to have in a well-designed experiment. Therefore, an experimenter faced with this set of calculations would seek a new research procedure with a higher value of 1-β (power).

Our example has illustrated the calculation of the power of a test when H_0 concerns the difference between μ's. The same general logic is used in

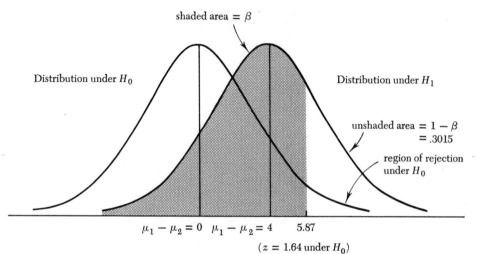

$(z = 1.64 \text{ under } H_0)$

Figure 9.2   Region of rejection for false H_0: $\mu_1 - \mu_2 = 0$ in the distribution H_1 in which $\mu_1 - \mu_2 = 4$ and $\sigma_{\overline{X}_1 - \overline{X}_2} = 3.58$, using one-tailed test and $\alpha = .05$.

determining the power of a test of other types of hypotheses. For example, in testing a hypothesis about μ we would use the z formula $(\overline{X} - \mu)/\sigma_{\overline{X}}$ and follow the same steps outlined above.

Factors Affecting Power of a Test

If you examine the z formula and the steps for determining the power of a test, you will see that the values of the mean and SD of the sampling distribution (in our example, $\mu_1 - \mu_2$ and $\sigma_{\overline{X}_1 - \overline{X}_2}$, respectively), as well as the value of z itself, all affect the probability figure we obtain. We will consider each of these terms and their implications for power one at a time. Again, we will use the two-population case as an illustration.

In our example the mean of the sampling distribution specified by the false H_0 is 0. The power of the test to detect this false hypothesis increases as the discrepancy between this mean and the true mean of the sampling difference (determined here by $\mu_1 - \mu_2$) increases. We can see this by examining Figure 9.2. If the $\mu_1 - \mu_2$ difference is small (close to the value specified by H_0), the true sampling distribution will show considerable overlap with the distribution under H_0. The value of β will therefore be high and $1 - \beta$, which defines the power of the test, correspondingly low. As $\mu_1 - \mu_2$ grows larger, the true sampling distribution is displaced further and further along the baseline and overlaps less and less with the distribution under H_0. Area β thus approaches a zero value and $1 - \beta$ approaches 1. As the difference between H_0 and the true state of affairs increases, then, the probability of rejecting the false H_0 approaches unity.

The second factor affecting power is the standard error of the sampling distribution, in our example $\sigma_{\bar{X}_1 - \bar{X}_2}$, which is equal to

$$\sqrt{\sigma_{\bar{X}_1}{}^2 + \sigma_{\bar{X}_2}{}^2} \quad \text{or} \quad \sqrt{\frac{\sigma_1{}^2}{N_1} + \frac{\sigma_2{}^2}{N_2}}$$

The basic components of the standard error formula, we can see, are N, the number of cases in the sample, and σ, the population SD. Consider first the implications of variations in N. As N increases, the value of each $\sigma_{\bar{X}}{}^2$ and hence of $\sigma_{\bar{X}_1 - \bar{X}_2}$ will *decrease*. Picture what would happen to the sampling distributions in Figure 9.2 as N increased from 10 per sample to larger numbers. The means of the distributions would remain the same, but with less variability (a smaller $\sigma_{\bar{X}_1 - \bar{X}_2}$) the distributions would each become taller and thinner. This, in turn, would reduce the overlap of the two sampling distributions. Reduction in overlap, we have already seen, increases $1 - \beta$, the power of the test. Large N's, then, make it less difficult to reject false hypotheses than small N's.

The second component affecting the standard error of sampling distributions is σ, with the standard error decreasing as σ decreases. In designing experiments, therefore, investigators try to minimize the influence of extraneous factors that may result in increased variability among their subjects and hence reduce the power of their test of significance.

The final factor affecting power is the α level we have chosen, which in turn determines z. We indicated earlier that the lower we set α, the less the probability of committing a Type I (α) error and the greater the probability of committing a Type II (β) error. Since power is defined by $1 - \beta$, this implies that the lower the α, the smaller the power of the test. We can demonstrate this by considering z. With a one-tailed test, as we saw in our example, the z associated with $\alpha = .05$ is 1.64. The value of $(\bar{X}_1 - \bar{X}_2)$ permitting us to reject H_0 was thus 1.64 ($\sigma_{\bar{X}_1 - \bar{X}_2}$) or 1.64(3.58) = 5.87. We determined that the probability of getting a difference of this size or greater in the true sampling distribution was .3015. The power of the test was thus .3015. If we had set α at .01, z would have been 2.33 and the critical value of $(\bar{X}_1 - \bar{X}_2)$ would have been 2.33(3.58) = 8.34. By finding the z-score equivalent of 8.34 in the true distribution [(8.34 − 4.00)/3.58 = 1.21], we determine from Table B that only 11.31 percent of the area falls beyond this point. We can see that the power of the test for $\alpha = .01$ is .1131, less than when we used the higher α level of .05.

The final factor affecting the power of a test is the nature of H_1. At a given α level the value of z is less when H_1 is unidirectional and evaluated by a one-tailed test than when a bidirectional H_1 and two-tailed test are used (for example, 1.64 vs. 1.96 for $\alpha = .05$; 2.33 vs. 2.58 for $\alpha = .01$). We have just seen that, other things being equal, the lower the value of this z the greater the power of the test. Unidirectional H_1's evaluated by one-tailed tests therefore are more powerful than bidirectional H_1's.

In a later chapter we will discuss still other techniques for testing hypotheses about the difference between population means. To understand these techniques we need to understand the concept of correlation, to which we turn next.

TERMS AND SYMBOLS TO REVIEW

Sampling distribution of differences between $\overline{X}$'s

t-ratio

t-table

Standard error of the difference between $\overline{X}$'s ($\sigma_{\overline{X}_1 - \overline{X}_2}$)

Assumptions underlying t

Type I (α) error

Estimated standard error of the difference ($s_{\overline{X}_1 - \overline{X}_2}$)

Type II (β) error

Power of a test ($1 - \beta$)

Null hypothesis about difference between μ's ($\mu_1 - \mu_2$)

Assumptions

Correlation

10

Scientific problems are usually attacked experimentally if possible. For example, we vary the intensity of background noise in which individuals are working in order to determine whether amount of noise affects performance. This is the customary procedure in experimentation; we vary a condition such as intensity of noise and note the result of that variation. Another example would be a study in which different groups of students are given different amounts or kinds of training in American history to find out whether such training affects their attitude toward the doctrine of states' rights.

Many phenomena, however, are not subject to experimental manipulation. We are unable either to change the movements of the planets or to control the birth rate in the United States to any great degree. To establish scientific laws about such phenomena we must use a correlational approach instead of an experimental one.

THE CORRELATION METHOD

Correlation, as the word itself suggests, is an interrelation between two or more sorts of conditions or events. We call these events _variables,_ because we are studying the way they vary. If families with many children tend to have a higher income than families with few children, we say there is a *positive correlation* between the two variables of family size and family income. If, however, families with many children tend to have a lower income than families with few children, we would say a *negative correlation* exists. The third possibility, that family income is the same regardless of family size, would lead us to say there is a *zero correlation* between family size and family income.

Hypothetical examples of these three classes of correlation (positive, negative, and zero) are shown graphically in the three parts of Figure 10.1. The graphs in the figure are called *scatter plots* because they show how the points scatter over the range of possible scores. Every point represents two values, an individual's score on one variable (X) and the same individual's score on a second variable (Y). Both variables are plotted on the same graph because we are interested in the relationship between them. As established in Chapter 3, the X variable is always measured along the horizontal axis and the Y variable along the vertical axis.

Figure 10.1 (a) shows the relationship between scores on a measure of self-esteem and a measure of assertiveness. This graph shows that degree of self-esteem is related to degree of assertiveness, individuals with high scores

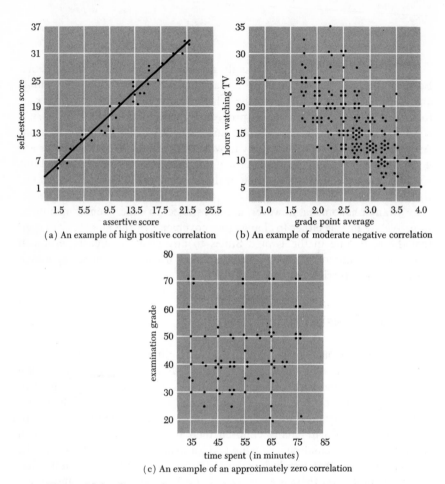

(a) An example of high positive correlation

(b) An example of moderate negative correlation

(c) An example of an approximately zero correlation

Figure 10.1 Scatter plots showing three types of correlation between X and *Y*

on one measure tending to have high scores on the other, and those with low scores on one tending to have low scores on the other. This, of course, is an example of a *positive* correlation. The next graph, Figure 10.1 (b), shows the relationship between high school students' grade point average and average amount of time per week spent watching television. In this scatter plot there is a tendency for students with *high* X scores to have *low* Y scores and vice versa. We therefore say that there is a *negative* correlation between X and Y; those who spend a lot of time watching TV tend to have a low grade point average and those with a good average tend to spend little time watching TV.

Figure 10.1 (c) shows a scatter plot of two variables with virtually no relationship between them. Each X score represents the time taken to complete an introductory psychology test, and the Y score represents the same person's grade on that test. Examination scores do not vary consistently in relation to the length of time spent by the students taking the examination. We describe this lack of relationship as reflecting an essentially *zero* correlation between X and Y.

We have just seen that the data of Figure 10.1 (a) and 10.1 (b) exhibit different *directions* of correlation—positive and negative, respectively. When we look at the graphs carefully, we note another difference. In Figure 10.1 (a) we have drawn a straight line in such a way that it falls close to as many points as possible. Most of the points do not lie far from the straight line, which indicates a very strong positive relationship between the two variables. If the relationship were perfect (that is, could not be improved upon), all the points would fall exactly along a straight line.[1] Similarly, all the points in a perfect negative correlation would fall along a straight line, except that high scores on the X measure would be paired with low scores on Y and vice versa. If we look at the negative relationship in Figure 10.1 (b), however, we observe that we could draw a straight line through the points that would describe them fairly well, but a number of points would lie at quite a distance from that line. The two variables, we see, are not as strongly correlated as in Figure 10.1 (a). In Figure 10.1 (c) no straight line would be particularly successful. A straight line going through the mean of the Y's and parallel to the horizontal axis or a straight line going through the mean of the X's parallel to the vertical axis would do better than a tilted line. The relationship between the variables is essentially zero. Thus, the second way in which scatter plots may differ is with respect to the amount or *degree* of correlation. Correlations may go from a perfect relationship, in which all the points lie on a straight line, to a zero relationship, in which the points are so scattered that no X shows a greater tendency to be paired with one Y more than any other and vice versa.

Correlation and Causation

In an experiment, we stated at the beginning of this chapter, we manipulate one variable to see if it affects another variable. For example, in a memory experiment we present each of a set of nonsense syllables for half a second and then five seconds later ask the subject to recall it. During the five-second interval one group is asked to read color names from a chart and the

---

[1] We actually are discussing here correlations that are *linear* in nature. That is, a straight line describes the points in the scatter plot better than a curved line. [See page 150 for a discussion of curvilinear correlation and Figure 10.2(b) for an example.] In a perfect *linear* correlation all points fall along a straight line.

other does nothing. We are interested in whether the color-naming activity during the interval influences the subjects' recall of the syllables.

The demonstration that two variables are correlated demonstrates *only* that they covary, that changes in the values of one tend to be associated with changes in the other. We cannot infer from the correlation alone that one of these variables has caused or brought about the other in the same way that we can in an experimental investigation. There are, of course, instances in which additional evidence suggests that variations in X do have some causal effect on Y. For example, the time since individuals last ate (X) is correlated with how hungry they report they are (Y). We know that hunger pangs are brought about by lack of food, so it is plausible to assume that X causes Y (but not the reverse: changing your report about how hungry you feel does not affect when you last ate). But, as this example illustrates, the causal path may not be direct or immediately obvious. A host of physiological events are taking place over time to affect the sensation of hunger. We do not know from the correlation, then, exactly how deprivation of food (X) brings about changes in reported hunger (Y).

While correlations may reflect some kind of causal link between X and Y, they more frequently occur because X and Y are both influenced by a common or a similar set of factors. The number of pairs of shoes a child has may be (negatively) correlated with the number of his brothers and sisters, but it is obvious that neither variable directly brings about the other. From our general knowledge we know that both are affected by a group of factors that centrally involve family income per family member.

We repeat our earlier statement. The mere demonstration that variables are correlated does not, in and of itself, tell us anything about causality; it shows us that variables tend to covary but not why. The "why" must be sought in additional kinds of data or in additional kinds of analyses of networks of correlations.

Unfortunately, there is often a temptation to infer a cause-effect relationship from no more than the correlation itself. A ludicrous example of the confusion resulting from an attempt to show a causal relation between two variables by correlational research was once given by Willoughby.[2] In attacking another scholar's argument that a high positive correlation between vocabulary and college grades meant that improvement in vocabulary would produce an improvement in grades, Willoughby stated that by the same reasoning a high positive correlation between the height of boys and the length of their trousers would mean that lengthening trousers would produce taller boys. In neither of these cases have we evidence proving that the manipulation of one variable directly controls the other.

[2] Willoughby, R. R. Cum hoc ergo propter hoc. *School and Society*, 1940, 51, 485.

Although correlation coefficients do not show what factors are responsible for the relationship between two variables, correlational techniques are nevertheless valuable for describing such relationships. The primary purpose of this chapter is to describe two of the most frequently used correlational techniques and to show how to make predictions of one variable from another when the correlation between the two is known. The latter technique enables us to do such things as predicting students' success in college from their entrance test scores or the adult heights of children from their heights at age six. Predictions of some accuracy are possible in these instances, because the pairs of variables are known to be correlated. Still another use of correlation, a modified *t* test for use with groups whose scores are correlated, will be described in Chapter 11.

The Pearson Product-Moment Correlation Coefficient (*r*)

Earlier in this chapter we developed in a general way the meaning of the terms positive and negative correlation and of relationships that are high, relatively low, or zero in magnitude. We also saw how, by inspecting a scatter plot, we often can specify the direction of the correlation between X and Y and make some kind of statement about its magnitude. Descriptions based on observing scatter plots, however, are not very precise. To overcome this deficiency we introduce a new concept, the *correlation coefficient*. This coefficient will have a specific numerical value for any given set of paired data. Positive values will correspond to what we have called positive correlation, negative values to negative correlation. Furthermore, high values of the coefficient, regardless of whether they are positive or negative, will correspond to what we have called high correlation, and low values to low correlation. What would be called zero correlation when seen on a scatter plot may not in fact have a correlation coefficient of exactly zero, although the value will surely be near zero.

In practice statisticians use several different correlation coefficients, depending on the type of data involved. In this chapter we will discuss two coefficients that describe the linear (straight line) correlation between variables. The technical name of the coefficient we will be using most often is the *Pearson product-moment correlation coefficient*, named in honor of Karl Pearson, one of the great pioneers of statistics, and symbolized as *r*. Because *product-moment correlation coefficient* is a highly technical term whose meaning is not essential to the research worker, we will not explain its origin. The three expressions—Pearson correlation coefficient, product-moment correlation coefficient, and *r*—will be used interchangeably in this chapter.

Basic Formula for *r*

We are now ready to define the Pearson correlation coefficient. The value of this coefficient (r) is equal to the mean of the products of the z-score for the X and Y pairs. This definition is stated algebraically below:

$$r = \frac{\sum z_X z_Y}{N} \tag{10.1}$$

where: $z_X = \dfrac{X - \bar{X}}{\tilde{\sigma}_X}$

$z_Y = \dfrac{Y - \bar{Y}}{\tilde{\sigma}_Y}$

N = the number of pairs

In determining r, we first calculate the z score for each X and for its paired Y. Next we multiply each person's z_X score by his z_Y score, giving us a ($z_X z_Y$) product for each pair of scores. Then we add up all these cross-products (that is, find $\sum z_X z_Y$) and divide by N. Now, the sum of any series of quantities divided by the number of quantities, we have seen, is the definition of the *mean* of these quantities. Thus, as we stated above, r is the mean of the z-score cross-products.

Now that we have a definition of r, it is appropriate to ask if the r values we obtain with this definition will have the same size and direction that our early statements about correlation suggest. We have said, for example, that a positive correlation will occur if relatively high X's and Y's tend to be paired and if relatively low X's and Y's tend to be paired. Can we say that r will be positive when these conditions are met? To answer this question, we will study how the value of r changes for different arrangements of z scores.

If X is larger than $\bar{X}$, z_X will be positive, because the formula $z_X = (X - \bar{X})/\tilde{\sigma}_X$ leads to positive z_X's in this case. Similarly, if X is smaller than $\bar{X}$, z_X will be negative. Corresponding statements will be true for Y. If we have a high positive correlation, high values on X quite uniformly go with high values on Y so that ($z_X z_Y$) will usually be the product of two positive numbers and will therefore be positive. Similarly, low values of X quite uniformly are paired with low values of Y in a positive correlation; the z scores of the low X and Y scores are both negative, so ($z_X z_Y$) again will be positive. Hence the sum of the cross-products ($\sum z_X z_Y$) will be positive when there is a high positive correlation. Therefore, r, which is the mean of the cross-products ($\sum z_X z_Y / N$), will also be positive. If we have a *low* positive correlation, most of the ($z_X z_Y$)'s will be positive, but the z's for quite a few pairs of scores will have opposite signs, so that their ($z_X z_Y$) value is negative. That is, in quite a few instances an individual will score above the mean on one variable (positive z score) and below the mean on the other (negative z score); the cross-product of these z's will be negative. When these negative and positive cross-products are added together, the overall sum will be positive but its magnitude will be less than when the correlation is high; hence, r will also be lower. Analogous reasoning could be used to show that a low negative correlation leads to an r with a negative value smaller than that of a high negative correlation. When there is absolutely *no* relationship between X and Y, negative and positive cross-products cancel each other out, so that both their sum and their mean are zero.

A perfect positive correlation takes the value of $+1.00$ and a perfect negative correlation the value of -1.00. In order to understand why this is so, recall that in the scatter diagram of a perfect linear correlation all the points lie along a straight line. When the correlation is perfect positive, this implies that the position of each paired X and Y in their respective distributions is the same: the individual who scores highest on X is highest on Y, the individual who scores at the mean of the X distribution also scores at the mean of Y, and so on. This in turn indicates that for each pair, $z_X = z_Y$, which can be expressed as z^2. When there is a perfect positive correlation, the formula for r therefore can be rewritten as $\sum z^2/N$. This expression is also the formula for the *variance* (SD^2) of a set of z scores. Thus, the square root of this expression gives us the SD of a set of z's. By definition, a z of 1.00 falls one SD unit from the mean of a distribution; the SD of a set of z scores is therefore 1.00. The variance is $(1.00)^2$, which also gives us 1.00. The value of r for a perfect positive correlation therefore is $\sum z^2/N = 1.00$. Parallel reasoning leads to the demonstration that a perfect negative correlation takes the value of -1.00. The value of r, then, ranges from $+1.00$ for a perfect positive correlation through zero for no correlation to -1.00 for a perfect negative correlation. This restriction of the range of possible coefficient values is extremely fortunate, because *it permits a direct comparison of* r*'s obtained from widely different sets of data without a correction for the size of the original X and Y values.* In this respect r is similar to z.

Two examples of the computation of r with the basic formula are presented in Table 10.1. In Case 1 most of the pairs of z's for X and Y are either both positive or both negative, producing positive values of ($z_X z_Y$) and a positive $\sum$ ($z_X z_Y$), which in turn produces a positive r of .873.[3] In Case 2 the z_X values are the same as those of Case 1, except that the z_X's that were positive have been changed to negative and those that were negative have been changed to positive. In other words, persons whose X scores were above $\overline{X}$ in Case 1 are now below $\overline{X}$ in Case 2, and persons whose X scores were below $\overline{X}$ in Case 1 are now above $\overline{X}$ in Case 2. Consequently, most of the $z_X z_Y$

---

[3] In this book no sign is attached to an r value unless it is negative. Consequently, an r value of .873 is positive and an r value of $-.873$ is negative.

Table 10.1   An example of the way *r* depends upon the pairing of *z* scores.

Case 1: Most $(z_X z_Y)$ products positive			Case 2: Most $(z_X z_Y)$ products negative		
z_X	z_Y	$(z_X z_Y)$	z_X	z_Y	$(z_X z_Y)$
1.4	1.3	1.82	-1.4	1.3	-1.82
1.9	1.3	2.47	-1.9	1.3	-2.47
1.0	1.1	1.10	-1.0	1.1	-1.10
.0	.5	.00	.0	.5	.00
$-.7$	.4	$-.28$	.7	.4	.28
$-.8$	$-.4$	.32	.8	$-.4$	$-.32$
$-.2$	$-.5$	.10	.2	$-.5$	$-.10$
$-.9$	-1.1	.99	.9	-1.1	$-.99$
$-.8$	-1.3	1.04	.8	-1.3	-1.04
$-.9$	-1.3	1.17	.9	-1.3	-1.17

$$\sum (z_X z_Y) + \text{'s} = +9.01 \qquad\qquad \sum (z_X z_Y) + \text{'s} = +.28$$

$$\sum (z_X z_Y) - \text{'s} = -.28 \qquad\qquad \sum (z_X z_Y) - \text{'s} = -9.01$$

$$\sum (z_X z_Y) = 8.73 \qquad\qquad \sum (z_X z_Y) = -8.73$$

$$r = \frac{\sum z_X z_Y}{N} \qquad\qquad\qquad r = \frac{\sum z_X z_Y}{N}$$

$$r = \frac{8.73}{10} \qquad\qquad\qquad r = \frac{-8.73}{10}$$

$$r = .873 \qquad\qquad\qquad r = -.873$$

products in Case 2 are negative, producing a negative *r* value of $-.873$. The size of *r* is the same as before, but its direction has been changed from positive to negative. This illustrates that the direction of correlation depends upon whether positive z_X's are paired with positive z_Y's and negative z_X's are paired with negative z_Y's or the reverse.

The Computational Formula for *r*

Despite the ease with which our basic formula for *r* can be remembered, an equation that permits raw scores (X and Y) to be used in place of *z* scores is more convenient for computing *r*. Of the several computational formulas available, a very useful version is shown below:

$$r = \frac{\sum XY - N\bar{X}\bar{Y}}{\sqrt{\sum X^2 - N\bar{X}^2}\ \sqrt{\sum Y^2 - N\bar{Y}^2}} \tag{10.2}$$

All of the components in the formula are familiar except for the expression $\sum XY$. This value is found by multiplying each X by the Y value paired with it and summing these XY products.

Table 10.2 shows how to use the computation formula to find r. The table shows the scores of ten students in a statistics class on their first and final examinations. The r for these two distributions is .363. We may conclude that there was a moderate tendency for students who did well in comparison to their classmates on the first exam to continue to do well on the final, and for students who initially did not do well to continue their poor performance on the final. We shall interpret this r of .363 in more detail as soon as we have discussed a few other properties of r.

Table 10.2   Demonstration of the use of the computational formula for r.

	First exam		Final exam		
Student	X	X²	Y	Y²	XY
1	31	961	31	961	961
2	23	529	29	841	667
3	41	1681	34	1156	1394
4	32	1024	35	1225	1120
5	29	841	25	625	725
6	33	1089	35	1225	1155
7	28	784	33	1089	924
8	31	961	42	1764	1302
9	31	961	31	961	961
10	33	1089	34	1156	1122

$$\sum X = 312 \quad \sum X^2 = 9920 \quad \sum Y = 329 \quad \sum Y^2 = 11003 \quad \sum XY = 10331$$

$$\bar{X} = \frac{\sum X}{N} = \frac{312}{10} = 31.2 \qquad \bar{X}^2 = 973.4$$

$$\bar{Y} = \frac{\sum Y}{N} = \frac{329}{10} = 32.9 \qquad \bar{Y}^2 = 1082.4$$

$$r = \frac{\sum XY - N\bar{X}\bar{Y}}{\sqrt{\sum X^2 - N\bar{X}^2} \sqrt{\sum Y^2 - N\bar{Y}^2}}$$

$$= \frac{10331 - 10(31.2)(32.9)}{\sqrt{9920 - 10(973.4)} \sqrt{11003 - 10(1082.4)}}$$

$$= \frac{10331 - 10264.8}{\sqrt{186} \sqrt{179}} = \frac{66.2}{13.64(13.38)} = \frac{66.2}{182.50}$$

$$= .363$$

WHAT *r* MEANS

The Pearson product-moment correlation coefficient for a particular set of data measures a specific type of relationship: the *linear correlation* between two variables. By this, we mean that *r* measures the degree to which a *straight line* relating X and Y can summarize the trend in a scatter plot. Figure 10.2 presents two such scatter plots, part (a) exhibiting a linear correlation and part (b) a nonlinear (or curvilinear) relationship between X and Y.

Although the data of Figure 10.2 (b) are best described by the dotted curved line of that figure, the straight line does come close to many of the data points. Thus, a straight line has some usefulness for summarizing the relationship between X and Y in this case.

What would happen if we computed *r*, a measure of linear correlation, for data that have a curvilinear relation? The answer is that we would be measuring how well those data approximate a linear relationship. For example, the curvilinear relationship shown in Figure 10.2 (b) is close enough to a linear one for us to expect a moderately high *r*. Many times, however, there may be a strong curvilinear relationship that is so different from a straight line that *r* is approximately zero. To illustrate, it might be found that a U-shaped curve best summarized the relationship between two variables. In such a case, the relationship would be so far from linear that *r* might be approximately zero. Consequently, when we interpret an *r* value, we must realize that we have an adequate measure of the *linear* correlation between two variables. This is true whether or not the underlying relationship is linear, but a low linear relationship does not necessarily imply a low curvi-

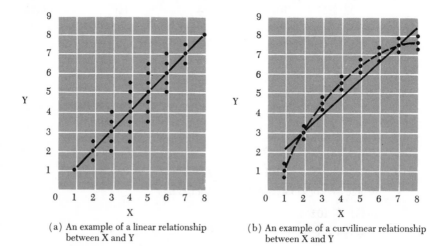

(a) An example of a linear relationship between X and Y

(b) An example of a curvilinear relationship between X and Y

Figure 10.2 Scatter plots of two hypothetical sets of data showing different kinds of relationships between X and *Y*

linear relationship as well. The degree of curvilinear relationship may be found by advanced statistical methods, which will not be considered here.

TESTING THE SIGNIFICANCE OF r

Our previous remarks about r have been appropriate only to the interpretation of the data for which the r was computed. For example, in Table 10.2 the data show a .363 linear relationship between students' scores on the first and final exams. This r does not permit us to make any statement about the relationship for a whole population of students, but it is a finding of some importance in itself.

In most instances, however, we wish to draw conclusions about a population of individuals; we wish to make inferences about the population r from the r of a sample. It might appear that we could test any null hypothesis that we wished about the population r, using procedures parallel to those described in Chapter 8 for testing hypotheses about population means and proportions. However, the hypothesis that is almost universally tested is that the population r is zero. We will therefore discuss only a test of this hypothesis.

Testing the null hypothesis that the population r is zero is quite simple. At any given α level the value of r required for significance turns out to be the same for all samples with the same df. (The df is given by N $-$ 2, one less than you might have expected; N, remember, is the number of *pairs* of scores.) Table D at the end of the book lists the significant values of r for the 5% and 1% levels. Whenever our obtained r is equal to or greater *in absolute size* than the tabled value, we can conclude that it is significant at the α level we have specified, that is, that the population r is not zero.

Now we can illustrate the procedure for testing the significance of r. Earlier we calculated the r between the first and final exam scores for students in a statistics class to be .363. Since this r was based on 10 pairs of measurements, there are 10 $-$ 2 or 8 df. Table D shows that for 8 df the smallest significant r at the 5% level is .632. Clearly, therefore, the obtained r of .363 is not significant. Consequently, we cannot reject the hypothesis at the 5% level that r is zero in the population from which the 10 students were drawn. We have no grounds on which to conclude that there is a linear correlation between the two sets of scores. What if we had data from 42 students and the r also *happened* to be .363? In this instance df is 42 $-$ 2 or 40. We see from Table D that the required value of r at the 5% level is .304. Since our obtained r is greater than this tabled value, we conclude in this instance that in the population as a whole the correlation *is* greater than zero in absolute value. This example shows us once more the importance of sample size in determining our conclusions about the null hypothesis, given sample data such as the value of r.

The Assumptions Involved in Testing *r* for Significance

Assumptions are merely conditions that must be met for a statistical test if the test is to be appropriate. For example, the *t* test depends upon the assumption or requirement that the experimental data are random samples from the populations being studied. Assumptions such as this are important because they are used in developing the statistical tests. Unless the assumptions of a test are met, then, the results of a test may be false and misleading.

Only two assumptions are required for testing the significance of *r*. First, the sample used must have been obtained by random sampling from the population concerned. Second, the population of X and Y scores must have a distribution that we will characterize by saying that X and Y must each be normally distributed, and that the relationship between X and Y must be linear. This is by no means a complete statement of the second assumption, but it suggests what to consider before testing *r* for significance. If the sample is randomly drawn, if X and Y are each normally distributed, and if the correlation between X and Y is linear, one may reasonably decide to perform this significance test.

THE USE OF *r* IN PREDICTION

Once a relationship between two variables has been found, it is often desirable to predict the value of one variable to be expected when a particular value of the other is obtained. For example, a university registrar might wish to predict college grades (Y) from performance on an aptitude test (X). If a method for such prediction were available, he could tell from the aptitude-test scores which applicants for admission would be most likely to succeed in college and which would be most likely to fail. Then he would be able to admit only those students with reasonable chances of success. This is exactly what is done in many colleges. Another example in which this type of prediction would be important is the case where some performance measure increases in a linear fashion with amount of practice. Through a study on the effects of practice in reading, it would be quite useful, if this were possible, to predict reading speed (Y) after different amounts of practice (X). This could serve two purposes: to indicate how much a specific person could improve his reading speed and to formulate a scientific law about the relationship between practice and reading speed.

To predict scores on one variable from scores on another (Y from X or X from Y), we first obtain pairs of data in which one X and one Y appear in each pair. Let us suppose we want to predict Y, given X. We then determine the best Y value to predict for each X value if the actual Y value were not known to us. The predicted Y's can be compared with the obtained Y values

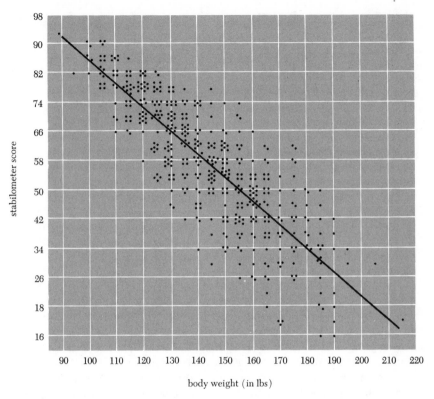

Figure 10.3   Scatter plot and regression line relating stabilometer scores to body weight

to indicate the success of the predictions. The ultimate usefulness of this technique will be in predicting unknown Y values when only X values are available.

Now that we know our goals in the prediction of one variable from another, we can turn our attention to the method of making such predictions. First, we recall that Figure 10.1 (a) was said to represent a high correlation because the straight line drawn through that graph came close to most of the plotted points. This line is a prediction line or, as it is often called, a *regression line*.

Another example of a regression line is given for the scatter plot presented in Figure 10.3, which shows the relationship between body weight (X) and ability to maintain bodily equilibrium (Y) observed with an instrument called a stabilometer.[4] This figure shows that subjects whose body

---

[4] Travis, R. C. An experimental analysis of dynamic and static equilibrium, *Journal of Experimental Psychology*, 1945, 35, 216–234.

weight is great have less ability to maintain equilibrium than those whose body weight is small. Thus, there is a negative correlation between body weight and ability to balance oneself, and the regression line goes from the top left of the graph to the bottom right. Notice that the regression line comes close to many data points, showing that good predictions can be made on the basis of negative correlations as well as positive ones.

Although the regression line shown in Figure 10.3 seems to conform to the relationship between the X and Y scores, other straight lines could be drawn that would seem just as satisfactory for predicting Y from X. In fact, if twenty people each drew a straight line through this figure to indicate where the regression line should be, twenty different regression lines might result. Therefore, we cannot trust ourselves to decide where a regression line should lie by simply looking at a graph. To make certain that everyone will make the same predictions of Y from X for any set of data we must have a procedure for finding the "best" regression line and the equation for that line.

The Equation of a Regression Line

The method that statisticians have adopted for finding the best-fitting regression line for predicting Y from X (or X from Y) is known as the *least-squares* method. This method attempts to specify the line that keeps at a minimum the squares of the deviations of all the points in the scatter diagram from the line (and hence keeps the sum of these squares at a minimum). The equation for the line yielded by this least-squares method is known as the *regression equation*. The regression equation for finding the predicted Y (Y_{pred}) from X is given below:

$$Y_{pred} = \left(\frac{rs_Y}{s_X}\right) X - \left(\frac{rs_Y}{s_X}\right) \bar{X} + \bar{Y} \qquad (10.3)$$

With this equation we can determine what value of Y to predict from any X value, provided that we know the values of r, s_Y, $\bar{X}$, and $\bar{Y}$.

The use of this equation may be illustrated by results of an actual experiment conducted to determine the effect of electric shock upon the drinking rate of laboratory animals (rats).[5] The amount of water each of twenty animals drank in a test period before shock is called X and the corresponding amount drunk in a test period after shock is called Y. The correlation between preshock and postshock drinking rate can be calculated to be .823. Other needed values are $\bar{X} = 5.77$, $\bar{Y} = 6.89$, $s_X = 1.107$, and $s_Y = 1.549$. We can develop the regression equation for predicting postshock rate

[5] Amsel, A., and Maltzman, I. The effect upon generalized drive strength of emotionality as inferred from the level of consummatory responses. *J. Experimental Psychology*, 1950, 40, 563–569.

(Y) from the preshock rate (X) by substituting these values into formula (10.3):

$$Y_{pred} = \frac{(.823)(1.549)}{(1.107)} X - \frac{(.823)(1.549)}{(1.107)} (5.77) + 6.89$$

After the appropriate multiplications and divisions have been performed, this equation reads:

$$Y_{pred} = 1.15X - 6.64 + 6.89$$

Finally, subtraction of 6.64 from 6.89 leads us to the final form for this regression equation:

$$Y_{pred} = 1.15X + .25$$

Table 10.3 shows the X for each of the experimental animals and the predicted Y for each X. It also presents the actual Y values in order to permit a comparison of Y_{pred} values with the actual Y's. (In practice, of course, we would predict Y from X only if Y were unknown.) This comparison indicates that the regression equation for these data leads to rather successful predictions. For example, when X = 3.3, the Y_{pred} is 4.05 and the actual Y is 3.60. When X = 8.0, then $Y_{pred} = 9.45$ and Y = 8.50.

A corresponding regression equation can also be found for predicting X from a given Y. This equation is:

$$X_{pred} = \left(\frac{rs_X}{s_Y}\right) Y - \left(\frac{rs_X}{s_Y}\right) \bar{Y} + \bar{X} \tag{10.4}$$

With a perfect r, you recall, all the points in the scatter diagram fall along a straight line. This straight line will also be the line generated by the two regression equations, one for predicting Y from X and the other for predicting X from Y. Thus, when r is perfect, the two formulas produce the same regression line. As r departs from ± 1.00, the two lines diverge until, when $r = 0$, they are at right angles to each other. When one wishes to predict Y from X when $r = 0$, the regression line starts at the value for $\bar{Y}$ on the Y axis and runs horizontally, parallel to the baseline or X axis. This line indicates that whatever the individual's X score, the best prediction is $\bar{Y}$, the overall mean for the Y variable. Knowing X, in short, does not improve our prediction of Y. Similarly, the regression line for predicting X from Y starts at $\bar{X}$ on the baseline and runs vertically, parallel to the Y axis. These relationships are illustrated in Figure 10.4.

Errors of Prediction. If r is perfect, use of the regression equation to predict Y from X or X from Y will yield perfectly accurate predictions.

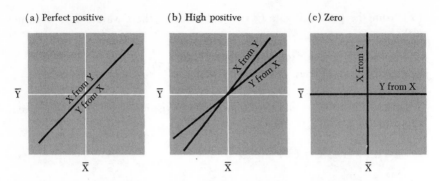

Figure 10.4 Relationship between the regression lines for predicting Y from X and X from Y when *r* is perfect positive, high positive, and zero

Table 10.3 An application of the regression equation in predicting postshock drinking (Y) from preshock drinking (X).

X	1.15X	Y_pred (1.15X + .25)	Y (actual)
5.0	5.75	6.00	7.00
6.0	6.90	7.15	5.80
6.5	7.48	7.73	7.80
4.7	5.40	5.65	5.90
5.8	6.67	6.92	6.80
7.4	8.51	8.76	8.00
5.0	5.75	6.00	4.00
5.2	5.98	6.23	6.40
4.5	5.18	5.43	5.10
5.8	6.67	6.92	7.70
5.6	6.44	6.69	5.80
8.0	9.20	9.45	8.50
6.3	7.24	7.49	9.00
5.5	6.32	6.57	7.60
6.9	7.94	8.19	8.40
5.2	5.98	6.23	6.60
3.3	3.80	4.05	3.60
5.5	6.32	6.57	7.30
5.8	6.67	6.92	7.00
7.4	8.51	8.76	9.50

$$\bar{X} = 5.77 \qquad \bar{Y} = 6.89$$

$$s_X = 1.107 \qquad s_Y = 1.549$$

$$r = .823$$

When r is less than perfect, errors can be expected. If you look at Table 10.3, for example, you observe that several Y_{pred} values are much in error, despite the high correlation between the two variables.

You may be tempted to use r as some kind of an accuracy indicator, concluding that an r of .90, for example, permits predictions that are 90 percent accurate and twice as accurate as predictions when $r = .45$. This use of r would be misleading. However, r^2 (called the *coefficient of determination*) does give some indication of accuracy. More precisely, r^2 tells us the proportion of the total variability among Y scores that can be accounted for by variability among X (and vice versa). When $r = .90$, for example, r^2 is .81; 81 percent of the variability among one set of measures can be accounted for by variations among the other. When r is .45, only 20.25 percent of the variability in one measure is explainable by variations in the other, so accuracy of prediction is considerably less.

Conversely, the amount of *unexplained* variability among a set of measures is given by $1 - r^2$. If $r = .50$, for example, $1 - r^2 = 1 - .25$; 75 percent of the variability in one variable is *not* explained by variability in the other, so prediction of X from Y or Y from X is far from perfect.

THE RANK-ORDER CORRELATION
COEFFICIENT (r_S)

Occasionally sets of data either are reported by their rank orders only, or it seems desirable to assign ranks to them and work with the ranks rather than the raw scores. Thus, a collection of rocks might be ranked in order from hardest to softest, even though no other measurement of hardness had been made first; or we might convert the performance of baseball teams, first reported as percentages of games won, to ranks by giving the team with the highest percentage rank number 1, the next team rank number 2, and so on. In these situations the r between the ranks themselves could be determined by means of our computational formula. However, a new formula will give exactly the same results with even less computational work. This formula, called the formula for the rank-order correlation coefficient (symbolized by r_S), is shown below. The symbol for the rank-order coefficient, you will note, differs from the one we used for the product-moment coefficient (r) by the addition of the subscript S.[6] The letter S has been chosen for this purpose in honor of Charles Spearman, who popularized the method.

$$r_S = 1 - \frac{6 \sum d^2}{N(N^2 - 1)} \tag{10.5}$$

[6] The rank-order coefficient may also be identified as Spearman rho (ρ). The symbol ρ, however, is usually reserved for a population value of the rank-order correlation coefficient and r_S for the value of the coefficient calculated from sample data.

where N is the number of pairs of ranks and d is the difference between a pair of *ranks* (never the difference between a pair of *scores*). You should notice that the 1 and the 6 in the formula are always used when r_S is to be found. They are constant regardless of the values of N or of d.

We may illustrate the application of this formula for the rank-order correlation coefficient by showing the computations required to find the correlation between two movie critics' rankings of ten movies that were released last year. Table 10.4 presents these computations. The ranks of one critic are listed as X and those of the other as Y. The difference, d, for each movie, is equal to X − Y. The value of r_S proves to be .636. This coefficient indicates a substantial agreement between the two critics.

Table 10.4 Computation of r_S for the rankings of ten movies by two movie critics.

Movie	(*Critic 1*) X	(*Critic 2*) Y	$d = X - Y$	d^2
A	1	6	−5	25
B	2	2	0	0
C	3	3	0	0
D	4	4	0	0
E	5	7	−2	4
F	6	1	5	25
G	7	5	2	4
H	8	9	−1	1
I	9	8	1	1
J	10	10	0	0
				$\sum d^2 = 60$

$$r_S = 1 - \frac{6 \sum d^2}{N(N^2 - 1)}$$

$$= 1 - \frac{6(60)}{10(99)} = 1 - \frac{360}{990} = .636$$

The Case of Tied Ranks

The rules for calculation of r_S must include what to do if there are ties for some ranks. What if, for example, we were correlating the final standings of the teams in the Big Ten Football Conference in the past two seasons and found that Iowa and Northwestern tied for sixth last year? We need to know how to treat their ranks before finding the correlation. A simple method is to assign each of the two tied teams the rank of 6.5, the average of the sixth and seventh ranks, since Iowa and Northwestern together must account for the ranks 6 and 7—the ranks *between* 5 and 8. Then r_S is computed as before.

We illustrate the assignment of values for the case of tied ranks by the following example, listing the intelligence quotients (IQ's) and ranks of ten persons whom we designate by the letters A through J:

	Persons									
	A	B	C	D	E	F	G	H	I	J
IQ	130	128	128	122	115	110	100	100	100	95
rank	1	2.5	2.5	4	5	6	8	8	8	10

First, notice that B and C tied for second place in their IQ's. Since B and C together account for the second and third persons in order of rank, 2 and 3 were averaged, giving 2.5 for B and 2.5 for C. Then D was assigned the fourth rank, since three persons were superior to him. Finally, G, H, and I were all assigned rank 8, because they were all tied and together represent the three ranks of 7, 8, and 9, which have an average of 8; and J, the last person, received a rank of 10 because nine persons were superior to him. (See footnote, page 234, for another approach to tied ranks.)

Interpreting the Rank-Order Correlation Coefficient

When r_S is found for data originally measured in ranks, it has the same value that r would have (except where ties have occurred) and may, therefore, be interpreted as a measure of the amount of linear correlation between ranks. If the data were originally expressed in units other than ranks (for example, in intelligence-test scores), r_S (which is, of course, the correlation between *ranks*) would *not* be equal to the r between the original scores. However, r_S and r will have similar values, so we are permitted to infer the *approximate* size of the r between the original variables from the size of r_S.

Despite the similarity between r_S and r, we cannot test the significance of r_S by means of Table D, the table of significant r's. Table D is inappropriate because the assumptions underlying r are not met with r_S. However, the significance of r_S may be tested another way. If we assume there is no relationship in the population between the ranks obtained on X and those obtained on Y (that is, $r_S = 0$), then there is a specific probability value that an r_S as large *in absolute value* as the one obtained in a particular sample could arise. If this probability is less than 5 percent, for example, we reject the hypothesis of no relationship at the 5 percent level. If this probability is greater than 5 percent, we accept the hypothesis.

Table E at the end of the book presents the values of r_S required for significance at the 5% level and at the 1% level. As in the case of r, the required values for significance depend upon the number of pairs, N. In Table E, we use N directly and do not have to find df. To illustrate the use of the table, suppose we have calculated an r_S of .514 between two sets of ranks obtained from a sample of twelve individuals. Can we conclude that r_S in the population from which the sample was drawn is greater than zero? When we refer to Table E, we see that with N = 12, an r_S of .591 is required for significance at the 5% level. Since our obtained r_S is below this value, we accept the hypothesis of no relationship between the two variables at the 5% level. Notice that in this case an r_S that seems quite large (.514) is not significantly different from zero at the 5% level because it is based upon a relatively small N.

We may summarize the facts about r_S by saying that it is a measure of the correlation between ranks and has the same value as would the r between ranks, assuming no ties in rank. Although r_S does not have the same value as the r between the scores on which the ranks were based, one would expect r_S and this r to have similar values. In the same way that Table D was used for r, Table E may be used to determine the minimum value of r_S that must be obtained if we are to conclude that the population value differs from zero at the 5% or 1% significance level.

TERMS AND SYMBOLS TO REVIEW

Positive, negative, and zero correlation

Degree of correlation

Scatter plot

Correlation vs. causation

Pearson product-moment correlation coefficient (r)

z cross-products

Perfect correlation

Prediction

Best-fitting regression line

Regression equation

Errors of prediction

Coefficient of determination (r^2)

Spearman rank-order correlation coefficient (r_S)

Tied ranks

The Design

of

Experiments

Independent
Random Samples
and Matched Samples

11

In Chapter 9 we learned how to use the t test to test hypotheses concerning the difference between population means. To repeat briefly, we measure a random sample from each of two populations on some characteristic and determine their $\overline{X}$'s. The null hypothesis we typically test is that the μ's of the populations from which the samples were drawn are equal, so that $\mu_1 - \mu_2 = 0$. From the results of the t test we state whether we will or will not reject the null hypothesis at a certain level of probability. If we reject the null hypothesis, we are saying that our $\overline{X}$'s come from populations whose μ's *do* differ with respect to this characteristic.

EXPERIMENTATION AND THE
t TEST

The t test is one of the most useful statistical methods available to the research worker. This is because much research takes the form of experiments in which the performance of at least two groups or samples of subjects is measured under *different conditions* to determine whether the conditions produce different performance. Thus, many investigations yield at least two mean scores—one from a group tested under one condition, the other from a group tested under another condition. If the t test indicates a significant difference between $\overline{X}$'s, the experimenter concludes that the two conditions *do* produce differences in performance. Note what this conclusion means: the investigator is assuming that the two groups would *not* have differed, except by sampling error, had they been measured under the *same* conditions. In other words, the experimenter assumes that if he had measured a large number of similar sets of two groups under the same conditions, he would have found no consistent difference between groups. If the assumption is true that the two groups of subjects would not have differed except for sampling error when measured under the same conditions, we call them *comparable groups*. Such groups may be used in research where their performance is measured under *different* conditions to find out whether the *conditions* produce significantly different performance.

From what we have just said it follows that a fundamental problem in research is the original selection of groups of subjects so that they are comparable at the beginning of the experiment. If groups are not comparable originally, conclusions drawn about the conditions studied may be in error regardless of whether the conditions had different effects. This chapter will describe the methods used to obtain comparable groups for research and present the appropriate statistics for each case.

Illustration of an Experiment

Before we take up the first method (random samples), we shall present an illustrative experiment. This will let us be a little more concrete about some of the aspects of an experiment mentioned earlier; we shall also have an illustration to which we can refer in order to clarify points to be taken up later.

Let us say an investigator wants to study the effect of distraction on learning. In the terminology used earlier, he is going to compare performance, as measured by speed of learning, under the condition of distraction with performance under the condition of no distraction. Each of his two groups of subjects consists of 20 undergraduates. The group that serves under the distraction condition is obtained in part from the 16 students in a senior psychology course taught by the experimenter, who requires them to serve. The experimenter then gets 24 volunteers from an introductory history course. After assigning 4 of these volunteers to the distraction group and the remaining 20 to the no-distraction group, he has his two samples.

The task used is the learning of a list of facts. Both groups learn the capital cities of thirty foreign nations. While learning the list, each member of the distraction group is subjected to hearing the tune "Johnson Rag" played through earphones. The no-distraction group learns the list without being subjected to sound.

At the conclusion of the study the investigator has, as a score for each subject, the number of times through the list (trials) required to learn it. He computes $\overline{X}$ for each group and finds a greater mean number of trials to learn for the distraction group. The difference between the $\overline{X}$'s is tested by the t test and found to be significant at better than the 5% level. The experimenter concludes that the sound was an effective distraction, producing slower learning in the distraction group.

We now ask: Shall we accept the investigator's conclusion? The answer is no, because it is entirely possible, even probable, that the groups were not comparable originally. A senior psychology class and an introductory history class are very likely different in many respects. For example, the group containing more history students may have known more foreign capitals to begin with. As we said earlier, *the groups in an experiment must be chosen so they would not differ except for sampling error if they were tested under the same conditions.* In this study this basic requirement probably was not met. The groups differed in too many ways for the research worker merely to assume, as he did, that they were alike at the outset on characteristics that could bias the results. There are other biasing characteristics (in addition to original knowledge of foreign capitals) on which the groups may well have differed originally. One such factor is ability to learn lists of specific facts; one group may have consisted, on the average, of faster learners. Another bias is that one group consisted of volunteers, whereas most of the other group were required to serve.

Now that we are prepared by this example of a poorly designed experiment, which no resort to a *t* test could save, let us take up an acceptable way of choosing groups for research—*random sampling*.

RANDOM SAMPLES

Earlier we referred to the groups in an experiment as *samples*. We now take up the case where the groups are *random samples*. As we already know, a random sample is one drawn such that every member of the population from which the sample is taken has an equal chance of being chosen. Still another requirement is that the selection of any given case does not affect the probability that any other case subsequently will be selected, so that each possible *sample* of a given size (each possible combination of N individuals) will also have an equal chance of being chosen. In an experiment involving two or more groups, we can achieve random samples if each subject is drawn randomly from the population and if the assigning of each subject to a group is also random.

How would random sampling work in the case of our illustrative experiment on distraction during learning? First, the experimenter would have to define the population. Let us say he defines it as all undergraduates, 1000 in number, in the college where he is working. (With a question so general as the effects of distraction, it is unlikely that the experimenter would be interested in such a circumscribed population, but the example will do for illustrative purposes.) Then the actual mechanics of drawing the two samples could be carried out in a number of ways—for example, by putting the name of each of the 1000 students on a slip of paper and drawing slips from a hat in which the slips have been thoroughly mixed. Strictly speaking, after a student's name has been selected, it should be returned to the population, so that on every draw each student has an equal chance, 1 in 1000, of being chosen. If replacement were not made, the remaining students' chances of being chosen would systematically increase as additional names were selected, $1/999$, $1/998$, $1/997$, and so on. However, this replacement procedure could result in the same individual's being selected for more than one sample or more than once for a single sample. Since, in a random-groups experiment, we cannot test the same subject twice, we usually ignore this requirement of random sampling and do not replace individuals once they have been selected. When the sample N is very much smaller than the population N, as it usually is, this departure from the strict requirements of random sampling has no serious consequences, and we proceed to analyze the data as if the requirement had been met.

Why Random Samples?

Recall once again the problem on which we are working: we want to obtain comparable groups of subjects for an experiment. How is this requirement met by random sampling? First, we remember that comparable groups were defined as those that do not differ *except for sampling errors.* Now, if we draw a *single* random sample from a population, the difference between the sample $\overline{X}$ and the population μ is sampling error. A large number of sampling errors is likely to be distributed normally. This means that positive and negative sampling errors would, in the long run, tend to cancel each other out; the μ of a normal distribution of random sampling errors approaches zero. It also means that any one sampling error is likely to be small, since in a normal distribution values close to μ occur more frequently.

If, now, we draw *two* random samples from the same population, differences *between the* $\overline{X}$'s are also sampling errors. Again, these sampling errors, which now are differences between $\overline{X}$'s of random samples, also tend to be distributed normally, or at least symmetrically, with the mean of the distribution approaching zero. Therefore, the probability is greater of drawing two random samples that differ by a small rather than a large sampling error. It follows that a research worker can use random samples from the same population for testing under the different conditions of his investigation, with confidence that any original difference between the samples is sampling error and is probably small.

There is another reason for basing research on random samples from the same population: the probability values given in Tables B and C, which are used to evaluate t ratios, are appropriate for random sampling. Recall that whenever we report that a difference between $\overline{X}$'s is significant at, say, the 1% level, we mean that the probability is no more than 1 in 100 that such a large difference would occur if the null hypothesis of no difference between population μ's were zero. Such statements are appropriate if random sampling and other assumptions of the t test are met.

The *t* Test with Independent Random Samples

In discussing the problem of obtaining comparable groups for research we have been considering the case where each subject needed is randomly and *independently* drawn from the population. Samples selected by this method are referred to as *independent random samples*. After an investigator measures the performance of his independent random samples under different conditions, he can test the null hypothesis he has set up about the difference between the population means.

We learned in Chapter 9 the necessary computations for the t test in this case. We wish here only to draw your attention again to the standard error of the difference ($s_{\bar{X}_1-\bar{X}_2}$), the denominator of the t ratio. When the samples are chosen randomly and the N's are equal, the $s_{\bar{X}_1-\bar{X}_2}$ is given by:

$$s_{\bar{X}_1-\bar{X}_2} = \sqrt{s_{\bar{X}_1}{}^2 + s_{\bar{X}_2}{}^2}$$

This is the formula for equal N's given in Chapter 9. We merely add here that it is the formula for $s_{\bar{X}_1-\bar{X}_2}$ to be used for *independent* random samples with equal N's; when we take up the case of matched samples, we shall find that another term must be added to the formula.

MATCHED SAMPLES

So far in our discussion of the problem of obtaining comparable groups for research we have considered only the case where each subject needed is randomly and independently drawn from the population. In some populations, however, the members occur in *pairs*. By this we mean that for each member of the population there is another member who is similar or in some way related. For example, in a population made up of sets of identical twins each individual is quite similar, on many characteristics, to the other member of the population who is his twin. Even a population of boys and their sisters can be considered for some purposes a population of pairs, since on some characteristics a boy will be more similar to his sister than he would be to a girl chosen at random.

As an alternative to selecting individuals by independent random sampling we may obtain comparable groups for research by sampling from a population of pairs of individuals. In this instance we do *not* draw each sample independently. Instead, we first draw a random sample of the *pairs*; that is, the two persons forming a pair are drawn simultaneously but randomly with respect to any other pair. We then form groups of subjects by picking randomly one member of each pair to be assigned to one group. The other member of each pair goes into the other group. Thus, if our population is composed of brothers and sisters, we would first select a random sample of pairs, each pair consisting of a boy and his sister. Then one member of each pair is picked *randomly* and assigned to one group; the other group is made up of the remaining member of each pair. The result will be that *each* group contains *both* boys and girls. Then, as usual, one group is tested under one condition, the other group under the other condition; we compare conditions or groups, not boys and girls.

Such random samples are called *matched* or *correlated* samples, and it will be worthwhile to examine how they differ from independent random samples. Let us first assume the drawing of a large number of sets of *independent random* samples (two samples in each set) from some defined population. If the samples are random and independent, there will *not* be a tendency for the $\overline{X}$'s of the two samples in a set to be similar. If $\overline{X}$ of one sample of a set is a relatively high value, $\overline{X}$ of the other sample is just as likely to be relatively low as high. We are saying that *independent random samples are not correlated*. Furthermore, if we plot a distribution of the *differences* between $\overline{X}$'s of sets of these random samples, we find that the *variability* of the distribution is relatively large because in some sets the $\overline{X}$'s are quite far apart. (When $\mu_1 = \mu_2$, for example, large discrepancies occur when one $\overline{X}$ falls relatively far above its μ and the other relatively far below.) In other words, the $s_{\overline{X}_1 - \overline{X}_2}$, the measure of variability of differences, is relatively large for $\overline{X}$'s of independent random samples.

Suppose, on the other hand, we have a population from which we can draw sets of *matched* samples (two samples per set), and again assume we draw a large number of sets of such samples. In this case we will find that the $\overline{X}$'s of the two samples of a set usually do show a tendency to be related in some way. If the pairs of scores are *positively correlated*, then when $\overline{X}$ of one sample of a set is a high value, $\overline{X}$ of the other sample also tends to be high. Similarly, low values tend to go together. For example, the IQ's of brothers and sisters are positively correlated. Suppose we draw sets of samples such that in each set one sample consists of boys and the other sample their sisters; that is, we draw matched samples. In a set where the boys have a high mean IQ, the mean IQ of their sisters will tend to be relatively high, and when the boys' $\overline{X}$ is low, the girls' $\overline{X}$ tends to be low. The result is that differences between $\overline{X}$'s of sets of positively correlated samples tend to be smaller than differences between $\overline{X}$'s of sets of independent random samples. *This means that the variability of a distribution of such differences—that is, the $s_{\overline{X}_1 - \overline{X}_2}$—will be less for positively correlated samples than for independent random samples.* Because of this, when we wish to test for the significance of a difference between $\overline{X}$'s of matched samples, we must use a different formula for the $s_{\overline{X}_1 - \overline{X}_2}$ than we use for random samples. Let us see what this formula is.

The t Test with Matched Samples

When an investigator has used matched samples in his research, the samples having been measured under different conditions, he will, as usual, have two distributions of scores at the conclusion of his research. As always, the distributions represent performance under the different experimental conditions, and the difference between the $\overline{X}$'s is to be evaluated by the t test.

For matched samples, the formula for $s_{\overline{X}_1 - \overline{X}_2}$ (again, the denominator of the t ratio) is:

$$s_{\overline{X}_1 - \overline{X}_2} = \sqrt{s_{\overline{X}_1}^2 + s_{\overline{X}_2}^2 - 2r_{12}s_{\overline{X}_1}s_{\overline{X}_2}} \qquad (11.1)$$

where: $s_{\overline{X}_1}$ = standard error of the mean of one group
$s_{\overline{X}_2}$ = standard error of the mean of the other group
r_{12} (read r sub one two) = the Pearson product-moment correlation between the scores of two groups—that is, between the two distributions of measures being tested for the significance of the difference.

Note that the initial part of the formula, up to the minus sign, is the same as the formula for random samples with equal N's and consists of the sum of the two $s_{\overline{X}}^2$'s. (Matched samples, you will also note, are made up of matched *pairs* and therefore always have equal N's.) For correlated samples the term $r_{12}s_{\overline{X}_1}s_{\overline{X}_2}$ is subtracted from this sum.

Let us consider the correlation coefficient r_{12} in that term. *It is the correlation between the two distributions, the $\overline{X}$'s of which are to be tested for significance.* Recall that to compute a Pearson r between two distributions we must be able to pair a score from one distribution with a score from the other. In any experiment using matched groups each pair of scores comes from the already paired subjects. Thus, all the terms under the radical sign in the $s_{\overline{X}_1 - \overline{X}_2}$ formula, the standard errors and the correlation, are obtained from the two distributions of scores resulting from the experiment. Either of these distributions may be identified by the subscript 1 and the other by the subscript 2.

What happens to the $s_{\overline{X}_1 - \overline{X}_2}$ for matched groups as r_{12} takes various values? When r_{12} is zero, the last term under the radical (everything after the minus sign) becomes zero, and we have the $s_{\overline{X}_1 - \overline{X}_2}$ formula for random groups. If there is no correlation between the groups, they are, in that respect, the same as independent random groups. If r_{12} is some positive value, the $s_{\overline{X}_1 - \overline{X}_2}$ will be *smaller* than when r_{12} is zero, and the higher the r_{12} the greater the reduction in the $s_{\overline{X}_1 - \overline{X}_2}$. Since a reduction in $s_{\overline{X}_1 - \overline{X}_2}$ leads to an increased t value with a given difference between two $\overline{X}$'s, we can see that this difference in $\overline{X}$'s is more likely to result in a significant t when divided by the $s_{\overline{X}_1 - \overline{X}_2}$ obtained from matched samples than from random samples. We also can see why there would be little point in using matched samples in an experiment unless the members of a pair originally were fairly similar. Unless the r_{12} is at least some medium to high value, there will be little reduction in $s_{\overline{X}_1 - \overline{X}_2}$.

Having computed the $s_{\overline{X}_1 - \overline{X}_2}$ for matched groups, we proceed to test the null hypothesis that the μ's are equal by obtaining t by the usual formula: $(\overline{X}_1 - \overline{X}_2)/s_{\overline{X}_1 - \overline{X}_2}$. In assessing the t value, the *df* that we use is N − 1, the number of *pairs* of measures minus one. Notice that *df* is less than if independent groups had been used, in which case df would have been $N_1 + N_2 - 2$. This means that for correlated samples a larger t will be required for significance at a given α level than if a t for independent samples had been calculated instead. When r is large enough (in a positive direction), the reduction in $s_{\bar{X}_1 - \bar{X}_2}$ that it brings about more than compensates for this problem, making it easier to find a significant difference with matched pairs when the null hypothesis is false.

Same Subjects in Both Groups

There is another case in which the $s_{\bar{X}_1 - \bar{X}_2}$ formula for matched groups is used—the case in which a *single* random sample of subjects serves in *both* conditions of the experiment.

Recall that our major concern, as we have harped so incessantly, is to obtain comparable groups for research. The two ways to obtain comparable groups that we have discussed—independent random samples and matched samples—both involve different individuals in the two groups. Most research requires that the groups contain different individuals, and this is the root of our problem; we must get different groups that can be considered comparable for the purposes of some experiment. This is not always easy to do, because people vary widely on almost every measurable characteristic. Therefore, it is especially pleasing to the research worker that in some studies he can use the same group of people in both conditions of the study. Whether or not he can use only one group depends upon the conditions being studied and the performance to be measured, but if he can, he measures the performance of a group once under one condition and again under the other condition. Thus, two scores are obtained for each subject, each score obtained under a different condition.

Whenever possible, it is advantageous to use the same subjects in both conditions. In the first place, the actual mechanics of getting the subjects and running them in the experiment are usually simpler, because only a single random sample is used. More importantly, there is no problem of obtaining comparable groups. The subjects serving in the different conditions are ideally matched; they are highly correlated because they are the same people. An individual is likely to be a better match for himself than is any other individual.

The Direct-Difference Method

We have discussed two cases, matched samples, and same subjects in both groups, where we must take account of the correlation between groups when we compute the $s_{\bar{X}_1 - \bar{X}_2}$ for a t test. Let us consider an experiment where such a correlation existed.

The example we have chosen is a case where the same subjects were used in both conditions of the study. The investigator wished to study the influence that children's physical attractiveness has on adults' judgment of their behavior.[1] He gave 20 adults a series of descriptions of relatively minor offenses, each supposedly committed by a real child. The subjects were asked to indicate on an objective rating scale how severe the punishment for each misbehavior ought to have been. A photograph of the purported miscreant accompanied each description, half of the photographs picturing children who had previously been judged to be physically attractive and the other half picturing children judged to be relatively unattractive. All 20 subjects therefore served in both conditions of the experiment. The raw data are presented in Table 11.1. The table shows the average severity-of-punishment rating made by each subject for the attractive and unattractive children.

We want to test the null hypothesis that the conditions had no differential effects—that is, that the μ's of the populations of scores for attractive and unattractive children are the same. For the t test we need the difference between $\overline{X}$'s and the $s_{\overline{X}_1 - \overline{X}_2}$. The formula for $s_{\overline{X}_1 - \overline{X}_2}$ for matched groups requires that we compute the $s_{\overline{X}}$ of each distribution, and the correlation (r_{12}) between the distributions. This is a lot of computation, so it is pleasant to state that there is a much shorter method of computing the t test for matched groups—one that yields the same t value. We shall call it the *direct-difference method*. This method can be used either with matched samples of different subjects or, as in our example, with same subjects in both groups.

To perform the t test for matched groups by the direct-difference method, we first obtain for each subject the *difference* between his two raw scores. This has been done in Table 11.1 in the column labeled *Difference*. In obtaining the difference score, it does not matter which raw score is subtracted from which as long as it is done in the same way for every subject. In Table 11.1 the attractive children's severity score is subtracted from the unattractive children's severity score; thus the difference score is negative whenever the attractive children's score is larger. The difference scores, for which we shall use the symbol D, can be considered and treated in the same way as a distribution of raw scores.

Once the distribution of direct differences, each difference having its correct algebraic sign, has been obtained, *all* further computations are based on it. First, we compute the algebraic mean of the differences ($\overline{X}_D$.) This is done by summing all the positive values, summing all the negative values, subtracting the sum of the negative values from the sum of the positive values to get $\sum D$, and then dividing $\sum D$ by N to get $\overline{X}_D$. In any matched-

---

[1] This hypothetical experiment is similar to an actual investigation performed by K. K. Dion (Physical attractiveness and evaluations of children's transgressions, *Journal of Personality and Social Psychology*, 1972, 24, 207–213).

Table 11.1   Severity-of-punishment scores for physically attractive and unattractive children.

Subject	Unattractive	Attractive	Difference	(Difference)²
1	15	17	−2	4
2	49	36	13	169
3	21	21	0	0
4	17	14	3	9
5	21	15	6	36
6	25	25	0	0
7	20	22	−2	4
8	56	38	18	324
9	16	19	−3	9
10	31	31	0	0
11	28	33	−5	25
12	44	39	5	25
13	35	29	6	36
14	48	41	7	49
15	32	31	1	1
16	37	27	10	100
17	45	21	24	576
18	29	27	2	4
19	28	32	−4	16
20	34	20	14	196

$$\sum D(+) = \overline{109}$$
$$\sum D(-) = -16$$
$$\sum D = \overline{93}$$

$$\bar{X}_1 = 31.55$$

$$\bar{X}_2 = 26.90$$

$$\bar{X}_D = \frac{\sum D}{N} = \frac{93}{20} = 4.65$$

$$\tilde{\sigma}_D{}^2 = \frac{\sum D^2}{N} - (\bar{X}_D)^2$$

$$s_{\bar{X}_D} = \sqrt{\frac{\tilde{\sigma}_D{}^2}{N - 1}}$$

$$= \frac{1583}{20} - (4.65)^2$$

$$= \sqrt{\frac{57.73}{20 - 1}} = \sqrt{3.03}$$

$$= 79.15 - 21.62 = 57.53$$

$$= 1.74$$

$$t = \frac{\bar{X}_D}{s_{\bar{X}_D}} = \frac{4.65}{1.74} = 2.67$$

$$df = N - 1 = 19$$

groups experiment N is always the number of *pairs* of raw scores, or the number of direct differences; thus, in this experiment N is 20.

The algebraic mean of the differences ($\overline{X}_D$) is *always* the same as the difference between the $\overline{X}$'s of the two raw-score distributions. Accordingly, in Table 11.1 $\overline{X}_D$ is 4.65 and identical with the difference between the two raw-score $\overline{X}$'s (31.55 and 26.90). The $\overline{X}_D$ is therefore the *numerator* in the *t* ratio; it is not necessary to compute the $\overline{X}$'s of the raw-score distributions, except as a check on accuracy of computation.

The $s_{\overline{X}_1 - \overline{X}_2}$ that is needed is obtained by computing *the standard error of the mean difference* ($s_{\overline{X}_D}$). Except for the fact that we are using difference scores we obtain the $s_{\overline{X}_D}$ in the same way as any standard error of the mean. First, we compute the SD^2 of the distribution of differences ($\tilde{\sigma}_D{}^2$). Again letting D stand for difference, we can write the SD^2 formula as:

$$\tilde{\sigma}_D{}^2 = \frac{\sum D^2}{N} - \overline{X}_D{}^2 \tag{11.2}$$

Then we compute $s_{\overline{X}_D}$ by the formula:

$$s_{\overline{X}_D} = \sqrt{\frac{\tilde{\sigma}_D{}^2}{N - 1}} \tag{11.3}$$

The standard error of the mean difference ($s_{\overline{X}_D}$) *is identical with the standard error of the difference that would be obtained by use of the long formula involving the correlation.* We can express the long method of doing a *t* test for matched groups, the direct-difference method, and the fact that the two methods give the same results by the following:

$$t = \frac{\overline{X}_D}{s_{\overline{X}_D}} = \frac{\overline{X}_1 - \overline{X}_2}{\sqrt{s_{\overline{X}_1}{}^2 + s_{\overline{X}_2}{}^2 - 2r_{12}s_{\overline{X}_1}s_{\overline{X}_2}}} \tag{11.4}$$

The form of the numerator of these formulas indicates, you will notice, that we are testing the usual null hypothesis that the two μ's are equal in value; therefore, we do not have to include the expression $\mu_1 - \mu_2$.

For the data in Table 11.1, $s_{\overline{X}_D} = 1.74$. Our *t* ratio is therefore 4.65/1.74, or 2.67. In any experiment using matched groups the number of degrees of freedom (*df*) for evaluating the *t* is N − 1, where again N is the number of pairs of measures. In our example there are twenty pairs of measures, so *df* is 19. With 19 *df* our *t* of 2.67 is significant at the 5% level. We conclude (with no great pleasure) that children who are physically attractive are judged less severely for their misbehavior than their less attractive peers.

The direct-difference method automatically takes into account the correlation that exists between the raw-score distributions (r_{12}), regardless of the size or algebraic sign of the correlation. Therefore, as we have said, the t value obtained by this method will be identical with the t value obtained by use of the long formula for the $s_{\bar{X}_1 - \bar{X}_2}$. We rarely use the long formula, which would probably save time only if we wanted to know the value of the correlation and thus already had computed r_{12}.

We recall once more that the r_{12} affects the *variability* of the distribution of direct differences. With a positive r_{12} the SD of these differences is reduced (as compared to the $\tilde{\sigma}_D$ if r_{12} is zero), and the higher the r_{12}, the smaller the $\tilde{\sigma}_D$. And, of course, the smaller the $\tilde{\sigma}_D$, the smaller the $s_{\bar{X}_D}$.

NONRANDOM SAMPLES

All the methods of obtaining comparable groups for research that we have discussed so far require random sampling from a defined population. Unfortunately, this requirement is an ideal that seldom can be met in practice. Much, perhaps most, research in the biological and social sciences is based on groups that were not randomly drawn from a defined population. In order to get any research done at all, the investigator may have to use whatever subjects are available to him. This procedure involves certain disadvantages that do not occur when random samples are used. We shall take up those matters later, but first let us describe some situations where random sampling is not feasible.

Randomized Groups Formed by the Investigator

It is often the case that a certain number of subjects are available, and the investigator has control over what subjects are assigned to what groups. For example, he might have a group of volunteers from his classes, or, if he is using animals, a group bought from a biological supply house. Such groups cannot be considered random samples from a defined population. Usually, however, the subjects *can* be assigned at random to the different conditions of the experiment. That is, half of the subjects can be assigned to one group and half to the other (in a two-group experiment) in such a manner that each individual, and each possible set of individuals, has an equal chance of being assigned to a particular group. To distinguish it from random sampling, this procedure is identified as *randomized* assignment of subject to group or *randomization*.

Strictly speaking, statistical tests based on the assumption of randomized assignment of subjects to groups should be used in analyzing the data from experiments using this type of design. However, statistical theory is far better developed for situations involving random sampling than for those involving randomization. Primarily for this reason the results of investigations using the randomization design are almost always, in practice, treated as though the groups had been obtained by random sampling and analyzed by statistical tests, such as t, based on the random-sampling assumption. Fortunately, the significance level given by t appears to be quite close to that yielded by a more exact test when both are applied to randomized data.

Intact Groups - Fixed

Sometimes subjects cannot be assigned at random to the groups, usually because the investigator has to use as his subjects individuals already organized into groups. Suppose, for example, he wants to compare two different methods of teaching introductory psychology. The only available subjects are the students taking the course. Suppose further that these students are divided into two sections meeting at different times. Very probably, the experimenter will have to use the sections as they exist, applying one method to one section, the other method to the other section.

In such a case the investigator must demonstrate that the groups are comparable for the purposes of the study. The problem of comparability is usually handled by comparing the groups' performance *before* introducing the different conditions. A pretest is given that measures the same kind of performance as the posttest (the measure that will be used to compare the different conditions). Thus, in our example of different methods of teaching psychology, the students could first be given an examination on the same kind of material that will be used at the end of the course to measure the effects of the different methods. The experimenter compares the mean scores of the two groups on this pretest, using the random-groups t test, and if the t value is not significant, he considers the groups sufficiently comparable for the purposes of the study.

What happens if, upon pretesting, the investigator finds a significant difference between groups? The common procedures are either to throw out some subjects, or to add further subjects, in order to make the groups comparable. If more subjects are available than are needed, as may be the case when the groups are formed by the experimenter, usually more subjects are added until comparable groups are achieved. With intact groups there may not be an extra supply of subjects. Comparable groups may be obtained in this case if enough of those subjects who got extremely high or low scores on the pretest are, in effect, eliminated by omission of their scores. We can also add subjects to intact groups by using other sets of such groups when they become available.[2]

---

[2] Students are referred to a discussion by D. T. Campbell and J. C. Stanley (*Experimental and Quasi-experimental Designs for Research*, Chicago, Rand McNally & Company, 1963) of the advantages and disadvantages of using intact groups and related topics.

Advantages of Random Sampling

One of the advantages of random sampling from a specified population is, as we have said, the assurance that differences between samples are not systematic or consistent differences. There is another important advantage. Since the basic requirement of random sampling (that every member of the population have an equal chance of being drawn) cannot be met without specifying the population, the limits within which an investigator can generalize his results are known. He can generalize to the population, and the sampling-error formulas enable him to state the confidence intervals for the population values.

In contrast, when an experimenter uses groups that are not random samples, he usually cannot specify the population. Therefore, he is in doubt about the generalizability of his results. For this reason, the research worker is very cautious in his generalizations. He makes no pretense that his results apply to any additional groups except perhaps those very similar to the ones he used. Although he may wish he could make predictions for larger groups, he will nevertheless have made a contribution if he shows whether or not the conditions he compared produced any difference in behavior, even if only in the groups he used. In short, much research still is needed to determine whether or not certain conditions have *any* effect on performance; such research may profitably proceed without precise specification of the population from which the samples were drawn.

TERMS AND SYMBOLS TO REVIEW

Comparable groups

Random samples

Matched samples

Direct-difference method

Mean difference ($\overline{X}_D$)

Estimated standard error of mean difference ($s_{\overline{X}_D}$)

Randomized samples

Intact groups

Simple
Analysis
of Variance

12

In previous chapters we have explained techniques for analyzing data from two different experimental conditions in order to test hypotheses about the effects of those conditions. We saw how to test the significance of differences between $\overline{X}$'s for two separate groups treated differently and also how to test the significance of differences between $\overline{X}$'s for the same group of subjects tested under two different conditions. As yet, however, we have paid no attention to the statistical analysis of findings from experiments in which the effects of three or more conditions are compared. This chapter will describe a technique called analysis of variance, one of the most widely used methods for making such analyses.

THE PURPOSE OF ANALYSIS OF VARIANCE

First of all, we ask why an experiment should ever include more than two conditions. Two answers may be given: (a) we may be interested in studying more than two conditions at a time, and (b) the data obtained with two conditions may give quite a different answer to our experimental problem than would comparable data from three or more conditions. Referring to our first point, suppose that an investigator is interested in the relative effectiveness of various methods of teaching French. We can see that there might be three, four, or even more teaching methods he would like to compare rather than only two. However (and this is our second point), if the experimenter studied only two teaching methods and found no differences in the results, he might conclude that teaching methods never differentially affect the students' performance, when, in fact, some other method that could have been included in an experiment with more than two conditions would have affected performance appreciably.

Let us consider another experiment in which more than two experimental conditions would be desirable. If we wished to determine the course of forgetting with the passage of time, we could train each of two groups of people to recite a list of twelve nonsense syllables and test one group for retention one hour after training and another group twenty-four hours after training. This would give us two points on the forgetting curve (the curve showing how much is remembered at various lengths of time after training), but it certainly would not give us enough information to permit us to plot the intermediate points on the curve or even to estimate the amount of forgetting after, say, six hours. Here too, we need more than two groups in order to obtain an adequate picture of the relationship between our experimental variable and the behavior being observed.

Once it is agreed that experiments involving more than two groups or conditions are sometimes desirable, we face the necessity of testing the significance of the differences among the $\overline{X}$'s obtained with the several conditions. Our first thought would surely be to compute t ratios to assess for each pair of groups the null hypothesis that the populations from which the groups were drawn had the same μ. We would thus test the significance of the difference between $\overline{X}$'s for all possible pairs of conditions. But this has certain disadvantages. For one thing, there may be a great number of pairs of conditions to be studied. If four methods of teaching French were to be compared, for example, there would be six different t ratios to be found.

Not only would a great deal of numerical work be required, but more important, the use of separate t tests for analyzing the results of an experiment based upon more than two groups would lead to results that could not easily be interpreted. If we obtain one significant t value out of six, what does this mean? Had we performed a single t test, we would have had .05 probability of obtaining a t significant at the 5% level by random sampling alone, but with six t tests our chances of having at least one of them significant because of random sampling is much larger. To carry this argument to its extreme, if a million t tests were involved in an experiment, a single significant t value would certainly be no indication that the conditions being varied had an effect.

We would not be handicapped by the fact that increasing the number of t ratios to be computed for an experiment increases the probability that a significant t will be found if we could state how this probability changes with the number of t-values to be computed. Unfortunately, this cannot readily be done, and so our use of t leads us to results we cannot evaluate easily.[1] Consequently, a single test for the significance of the differences among all the $\overline{X}$'s of the experiment seems necessary as a means of permitting a more correct determination of the significance of differences obtained in the experiment. This chapter will introduce you to one such test, the *analysis of variance* technique. This technique is appropriate for use with two or more groups. However, it is most frequently employed with three or more groups, since the t test is available for experiments involving two groups. First we will describe the principles of analysis of variance; then we will show how to apply that method to scientific data.

---

[1] If each of the t ratios had been computed from a completely different set of data, we could employ the tables presented in Wilkinson, B. A statistical consideration in psychological research. *Psychological Bulletin*, 1951, 48, 156–158. However, when all possible pairs of conditions in a single experiment are compared, the same $\overline{X}$ and SD values are used several times, and the t ratios are not independent of each other.

THE ABC's OF ANALYSIS OF VARIANCE

This section will describe the analysis of variance technique in very general terms. The basic underlying fact is that the total variability of a set of scores from several groups may be divided into two or more categories. In the type of experimental design we are considering there are two categories of variability: the variability of subjects within each group and the variability between the different groups. To visualize these two kinds of variability, imagine that we have four experimental conditions, each of which is applied to a different group. Suppose each condition yields widely different scores, so that members of Group 1 all have smaller scores than members of Group 2, who in turn have smaller scores than members of Group 3, and so on. This gives us the situation graphed in Figure 12.1. Notice that $\overline{X}$ values are given for each group separately and also for the set of four groups combined, $\overline{X}$ in the latter case being called $\overline{X}_{tot}$ ($\overline{X}$ of total). Now we can identify the variability observed in the experiment: the variability of the scores from $\overline{X}_{tot}$ is called the *total variability*, the variability of the scores from their group $\overline{X}$'s is called the *variability within groups*, and the variability of the group $\overline{X}$'s from $\overline{X}_{tot}$ is called the *variability between groups*.

Consider, for example, the people who have the score marked X_2. The deviation $(X_2 - \overline{X}_{tot})$ contributes to the total variability, but that deviation is composed of the two quantities $(X_2 - \overline{X}_2)$ and $(\overline{X}_2 - \overline{X}_{tot})$, as Figure 12.1 makes clear. Since $(X_2 - \overline{X}_2)$ is the difference between a score and its group $\overline{X}$, and $(\overline{X}_2 - \overline{X}_{tot})$ is the difference between the group $\overline{X}$ and $\overline{X}_{tot}$, the contribution of X_2 to the total variability is made up of one

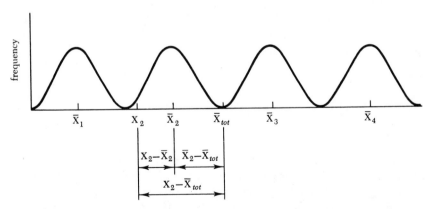

Figure 12.1 An example of four groups of scores illustrating variability within groups and variability between groups.

part that is variability within a group and another that is variability between groups. Since any score's deviation from $\overline{X}_{tot}$ may be divided algebraically into these two parts, it is possible to express the overall variability observed in an experiment as the sum of these two types of variability—the sum of the variability between groups and the variability within groups.

Once we have assimilated the idea that one kind of variability in an experiment is between groups and another kind is within groups, we can let our previous experience with statistical tests suggest how this idea can be employed in analyzing differences in $\overline{X}$ values obtained from several groups. Thinking back to our study of t (and our test of the null hypothesis that $\mu_1 - \mu_2 = 0$), we remember that the numerator of the t ratio is equal to $\overline{X}_1 - \overline{X}_2$ and that the denominator is $s_{\overline{X}_1 - \overline{X}_2}$. Since $\overline{X}_1 - \overline{X}_2$ is a measurement of the difference between groups, it could be called an expression of the variability between groups. Similarly, $s_{\overline{X}_1 - \overline{X}_2}$ could be considered a measurement of the variability within groups, making t the ratio of variability between groups to variability within groups. In that case we could say that a significant t occurs only when the ratio of between-groups variability to within-groups variability is as large as the significant t's presented in Table C. Applying the same reasoning to experiments involving more than two groups, we first assume the null hypothesis that the μ's of the groups are all the same ($\mu_1 = \mu_2 = \mu_3 = \cdots = \mu_k$), so that the overall difference is zero. We then obtain a ratio of these two types of variability and, provided this ratio is sufficiently large, we call the differences among $\overline{X}$'s significant. This ratio, we shall discover shortly, is identified by the letter F. The analysis of variance technique is simply a method which provides an objective criterion for deciding whether the variability between groups is large enough in comparison with the variability within groups to justify the inference that the means of the populations from which the different groups were drawn are not all the same.

The relationship between the t and the F ratio is more formal than the parallel we have just described. In fact, when only two groups are being compared, F has the value of t^2. Further, when one is significant (or nonsignificant) at a given α level, so is the other.

THE CONCEPT OF SUMS OF SQUARES

Basic Ideas

The cornerstone in the analysis of variance technique is the *sum of squares*, an old concept masquerading under a new name. *The sum of squares is simply $\sum x^2$, the sum of squared deviations from the mean*, an expression we used earlier in defining the variance, or SD^2. (We remember that $SD = \sum x^2/N$, where x is the deviation of the score X from the mean.)

When the value of x is large—that is, when a score (X) is far from $\overline{X}$—x^2 is also large. Thus a large $\sum x^2$ (sum of squares) occurs when the scores tend to be widely dispersed about $\overline{X}$. But this is also what we mean by great variability. Consequently, a large value for the sum of the squares indicates a large amount of variability; small sums of squares values indicate small amounts of variability.

Types of Sums of Squares Values

We have already stated that the total variability in an experiment may be divided into two parts: the variability of subjects within groups, and the variability between different groups. Now we wish to find ways of computing sums of squares that will correspond to the total variability and to its two parts. These three sums of squares, which we shall call, respectively, total sum of squares (SS_{tot}), within-groups sum of squares (SS_{wg}), and between-groups sum of squares (SS_{bg}), will be discussed presently. First, however, we must list some new symbols that will be used in the remainder of this chapter:

$\overline{X}_{tot}$ = the mean of all scores in the experiment (also called the grand mean or the overall mean).

$\overline{X}_g$ = a general expression for the mean of the scores in any group (g).

N_{tot} = the number of scores in the experiment.

N_g = a general expression for the number of scores in any group.

k = the number of groups.

$\sum_{tot} X$ = the sum of all the scores in the experiment.

$\sum_{tot} X^2$ = the sum of all the squared scores in the experiment.

We also need to distinguish between the scores of the different groups. A score from Group 1 is indicated by X_1, a score from Group 2 by X_2, a score from Group 3 by X_3, and so forth. This numerical subscript is similarly used to identify the mean and N of each group (for example, $\overline{X}_4$ is the mean of Group 4 and N_4 is the number of cases in that group).

The Definition of SS_{tot}. The total sum of squares (SS_{tot}) is just what our previous statement would suggest: *the sum of the squared deviations of every score from the grand mean* ($\overline{X}_{tot}$) *of all the scores in the experiment.* This definition may also be expressed in an equation:

$$SS_{tot} = \sum (X - \overline{X}_{tot})^2 \tag{12.1}$$

Because SS_{tot} is to represent all the variability recorded in an experiment, we must find $(X - \overline{X}_{tot})^2$ for each X in the entire experiment and then sum all of these $(X - \overline{X}_{tot})^2$ quantities. Accordingly, the number of squared deviations to be added will be equal to the total number of subjects in the experiment (N_{tot}). To repeat, we subtract $\overline{X}_{tot}$ from each person's score (X), square this deviation, and then sum these squared deviations.

The Definition of SS_{wg}. The sum of squares within groups is most easily understood as a combination of sums of squares within separate groups. For example, we define the sum of squares within Group 1 (SS_{wg1}) by the following equation:

$$SS_{wg1} = \sum (X_1 - \overline{X}_1)^2$$

(Note that X_1 is a score in Group 1, and that we must find $(X_1 - \overline{X}_1)^2$ for every X_1 in Group 1.) This is an expression of the variability within Group 1. You can see that it serves the same purpose for Group 1 that SS_{tot} does for the entire experiment.

We define the sum of squares within Group 2 (SS_{wg2}) by writing the preceding equation in terms of X_2 and $\overline{X}_2$ rather than X_1 and $\overline{X}_1$. A similar change is made for Group 3 and for any other group in the experiment. Once the sum of squares within each of the groups is known, we add them, because we want only one SS_{wg} expression. Accordingly, we define SS_{wg} for the entire experiment as the sum of the SS_{wg}'s for the separate groups. Writing this definition in terms of the quantities that make up the individual SS_{wg}'s, we have:

$$SS_{wg} = SS_{wg1} + SS_{wg2} + SS_{wg3} \tag{12.2}$$

or

$$SS_{wg} = \sum (X_1 - \overline{X}_1)^2 + \sum (X_2 - \overline{X}_2)^2 + \sum (X_3 - \overline{X}_3)^2$$

for the case of three groups. With a different number of groups (k) the number of terms on the right-hand side of the equation would equal the new value of k. For example, with five groups (when $k = 5$) SS_{wg} would be defined as $SS_{wg} = SS_{wg1} + SS_{wg2} + SS_{wg3} + SS_{wg4} + SS_{wg5}$.

The Definition of SS_{bg}. From the above we would expect that SS_{bg} would be based upon the squared deviation of each $\overline{X}_g$ from $\overline{X}_{tot}$, just as other SS's are based upon deviations of scores from $\overline{X}$'s. This is exactly what is done. First we find $(\overline{X}_g - \overline{X}_{tot})^2$ for each $\overline{X}_g$. Then we weight each $(\overline{X}_g - \overline{X}_{tot})^2$ quantity with the number of cases in its group (N_g) and add the weighted quantities $N_g(\overline{X}_g - \overline{X}_{tot})^2$ to find SS_{bg}. In general, then, we can say the following:

$$SS_{bg} = \sum N_g(\overline{X}_g - \overline{X}_{tot})^2 \tag{12.3}$$

Taking a specific number of groups ($k = 3$), we can write this equation in a new form:

$$SS_{bg} = N_1(\overline{X}_1 - \overline{X}_{tot})^2 + N_2(\overline{X}_2 - \overline{X}_{tot})^2 + N_3(\overline{X}_3 - \overline{X}_{tot})^2$$

since N_g is N_1 for the first group, N_2 for the second group, and N_3 for the third group and $\overline{X}_g$ has comparable values.

Relationship Between SS_{tot} and SS_{wg}

We have said that SS_{wg} typically accounts for part, but not all, of SS_{tot}. For instance, if we compared the mathematical-aptitude scores of physics and English majors, the variability of scores within the two groups would account for part of the total variability (SS_{tot}). This is not the sole source of variability, however; we would expect the physics majors to exhibit greater mathematical aptitude than the English majors, and this difference between groups would be reflected in an SS_{bg} value, which would also be part of SS_{tot}.

In some other experiments there might be no difference, or almost none, between the groups. For example, three teaching methods—a lecture method, a discussion method, and a project method—used with three different groups of students of freshman history might result in identical $\overline{X}$'s on their exams. Then each $\overline{X}_g$ would have the same value as $\overline{X}_{tot}$, and there would be no SS_{bg}. In such a case all of SS_{tot} would be accounted for by SS_{wg}. In the next three paragraphs this fact, that SS_{wg} is *part* of SS_{tot} when there are differences in the $\overline{X}$'s of the groups, and that SS_{wg} is *all* of SS_{tot} when there are no differences in the $\overline{X}$'s of the groups, is elaborated for the case of three groups.

If all groups have the same mean, then $\overline{X}_1 = \overline{X}_2 = \overline{X}_3$. Furthermore, each $\overline{X}_g$ (group $\overline{X}$) is equal to $\overline{X}_{tot}$. Consequently, we may substitute $(X - \overline{X}_{tot})$ for each $(X - \overline{X}_g)$ value in our definition of SS_{wg}. But $(X - \overline{X}_{tot})$ is the basic component in the computation of SS_{tot}. Therefore, SS_{tot} will now equal SS_{wg}, as the set of balances in Figure 12.2 shows.

Figure 12.2 Demonstration that SS_{tot} is equal to SS_{wg} when the group $\overline{X}$'s are equal to $\overline{X}_{tot}$

The reason that SS_{wg} balances SS_{tot} is suggested by the objects in the scale. Because every X must come from some group, it is either an X_1, an X_2, or an X_3 in this situation. Therefore, the quantity on the right of the balance is exactly the same as that on the left; on each side of the balance we are subtracting $\overline{X}_{tot}$ from each score separately, squaring the differences and summing them. Accordingly, $SS_{tot} = SS_{wg}$ when all $\overline{X}_g$'s are equal, as in the case of the three teaching methods that proved equally effective.

Of course, the situation above is an exceptionally rare one. Even when three experimental conditions have identical effects, the X_g's usually differ slightly from one another. In this case the $\overline{X}_g$'s are not equal to $\overline{X}_{tot}$, SS_{tot} is greater than SS_{wg}, and the scale no longer balances, as illustrated in Figure 12.3.

$$\Sigma (X_1 - \overline{X}_1)^2 +$$
$$\Sigma (X_2 - \overline{X}_2)^2 +$$
$$\Sigma (X_3 - \overline{X}_3)^2$$

Figure 12.3 Demonstration that SS_{tot} is greater than SS_{wg} when the group $\overline{X}$'s are not equal to $\overline{X}_{tot}$

$$\Sigma (X - \overline{X}_{tot})^2$$

SS_{tot} SS_{wg}

We can put the scale in balance again if we add an appropriate amount to the right-hand side. This amount will always be the value of SS_{bg}, as we have indicated in Figure 12.4. However, SS_{bg} is not simply a convenient quantity dreamed up to account for the difference between SS_{tot} and SS_{wg}. It is a meaningful measurement of the difference between group $\overline{X}$'s and will prove extremely useful to us.

We now know the definitions of SS_{tot} and its two components, SS_{bg} and SS_{wg}. We also understand how the differences among $\overline{X}_g$'s affect these SS values. Before we can apply our knowledge in a practical way, however, we must know how to compute the values for SS_{tot}, SS_{bg}, and SS_{wg}.

Figure 12.4 Demonstration that $SS_{tot} = SS_{wg} + SS_{bg}$ when the group $\overline{X}$'s are not equal to $\overline{X}_{tot}$

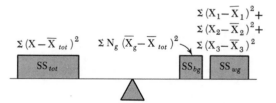

$$\Sigma (X - \overline{X}_{tot})^2 \qquad \Sigma N_g (\overline{X}_g - \overline{X}_{tot})^2$$

$$\Sigma (X_1 - \overline{X}_1)^2 +$$
$$\Sigma (X_2 - \overline{X}_2)^2 +$$
$$\Sigma (X_3 - \overline{X}_3)^2$$

SS_{tot} SS_{bg} SS_{wg}

Computational Formulas for Finding
Sums of Squares

Application of the basic formulas that we have just discussed for finding sums of squares is a long, laborious process that involves determining the deviation of each score from the overall or grand mean, the deviation of each score from its own mean, squaring each of these sets of deviations, and so on. Just as with SD and r we can simplify our work in calculating SS values by using special computational formulas, which lead to precisely the same answers as do the basic formulas but are usually faster. Although these new equations can be derived algebraically from the basic definitions, we will present them without proof. Following this presentation, we will demonstrate their application to a set of data.

First, SS_{tot} is found by the computational formula:

$$SS_{tot} = \left(\sum_{tot} X^2 \right) - \frac{(\sum\limits_{tot} X)^2}{N_{tot}} \tag{12.4}$$

The formula states that SS_{tot} is computed by first squaring each X and adding all the X^2 values together, giving us $\sum\limits_{tot} X^2$. Next the sum of all the X values is found; this *sum* is squared and then divided by the total N, giving us $(\sum\limits_{tot} X)^2/N$. Finally, this latter result is subtracted from $\sum\limits_{tot} X^2$ to give SS_{tot}.

Next, SS_{bg} is found by the formula:

$$SS_{bg} = \sum_{g} \left[\frac{(\sum X_g)^2}{N_g} \right] - \frac{(\sum\limits_{tot} X)^2}{N_{tot}} \tag{12.5}$$

Thus, we determine SS_{bg} by first finding for each group its $\sum X$, squaring it, and dividing the result by the corresponding N. The resulting quantities from all the groups are added together (as indicated by the symbol $\sum\limits_{g}$), and the quantity obtained by dividing $(\sum\limits_{tot} X)^2$ by N_{tot} subtracted from this sum. (You will note that the latter quantity was also used in determining SS_{tot}, so it needs to be calculated only once.) The computational formula for SS_{bg} could also be written for k groups as follows:

$$SS_{bg} = \left[\frac{(\sum X_1)^2}{N_1} + \frac{(\sum X_2)^2}{N_2} + \cdots + \frac{(\sum X_k)^2}{N_k} \right] - \frac{(\sum\limits_{tot} X)^2}{N_{tot}}$$

Finally, since we know that SS_{tot} is composed of SS_{bg} and SS_{wg} we can find SS_{wg} by subtraction:

$$SS_{wg} = SS_{tot} - SS_{bg} \qquad (12.6)$$

We could also compute SS_{wg} directly, instead of by subtraction, by calculating SS_{wg} for each group separately and then adding these values together to obtain the overall SS_{wg}. These operations are expressed by the following formula:

$$SS_{wg} = \sum_g \left[(\sum X_g^2) - \frac{(\sum X_g)^2}{N_g} \right] \qquad (12.7)$$

With the computational formulas for SS_{tot}, SS_{bg}, and SS_{wg} before us, we can illustrate their use with a specific set of scores. Table 12.1 presents the results of a hypothetical experiment on the effects of courses in music appreciation upon attitudes toward classical music. Group 1 members took no courses in music appreciation, Group 2 took one course, and Group 3 took two courses before having their attitudes measured. Our first step, as shown in Table 12.1, is to find the $\sum X$ and $\sum X^2$ for each group and then to sum each of these quantities across groups to obtain $\sum_{tot} X$ and $\sum_{tot} X^2$. We then compute SS_{tot}, SS_{bg}, and SS_{wg} using our computational formulas. You will notice that we have obtained the value of SS_{wg} twice, first by subtracting SS_{bg} from SS_{tot} and then by computing it directly. The fact that the two procedures yield the same value for SS_{bg} not only shows that our calculations were accurate but also demonstrates our earlier statement that $SS_{bg} + SS_{wg}$ always equals SS_{tot}.

One more fact about the computation of SS values is important to us. If we look carefully at the basic formulas for SS values, we see that no SS can ever have a negative value. This is not obvious from the computational formulas, but must also be true for them. Consequently, a negative SS value is always a sign that a mistake in arithmetic has been made and that recalculations are necessary.

THE CONCEPT OF VARIANCE OR MEAN SQUARE

The use of sums of squares as measurements of variability has only one disadvantage: the size of SS depends very much upon the number of measurements upon which it is based. Other things being equal, a sample of 100 measurements from a given population will yield about twice as large an SS as a sample of 50 from that population, because there will be twice as many

Table 12.1 Application of the computational formulas for SS_{tot}, SS_{bg}, and SS_{wg} to the data from the hypothetical attitude experiment.

X_1	$X_1{}^2$	X_2	$X_2{}^2$	X_3	$X_3{}^2$
68	4624	78	6084	94	8836
63	3969	69	4761	82	6724
58	3364	58	3364	73	5329
51	2601	57	3249	67	4489
41	1681	53	2809	66	4356
40	1600	52	2704	62	3844
34	1156	48	2304	60	3600
27	729	46	2116	54	2916
20	400	42	1764	50	2500
18	324	27	729	32	1024
420	20448	530	29884	640	43618

$$\sum_{tot} X = \sum X_1 + \sum X_2 + \sum X_3 = 420 + 530 + 640 = 1590$$

$$\sum_{tot} X^2 = \sum X_1{}^2 + \sum X_2{}^2 + \sum X_3{}^2 = 20448 + 29884 + 43618 = 93950$$

$$N_{tot} = N_1 + N_2 + N_3 = 10 + 10 + 10 = 30$$

$$SS_{tot} = \left(\sum_{tot} X^2\right) - \frac{(\sum_{tot} X)^2}{N_{tot}}$$

$$= 93950 - \frac{(1590)^2}{30}$$

$$= 93950 - 84270$$

$$= 9680$$

$$SS_{wg} = SS_{tot} - SS_{bg}$$

$$= 9680 - 2420$$

$$= 7260$$

or

$$SS_{bg} = \sum_g \left[\frac{(\sum X_g)^2}{N_g}\right] - \frac{(\sum_{tot} X)^2}{N_{tot}}$$

$$= \frac{(\sum X_1)^2}{N_1} + \frac{(\sum X_2)^2}{N_2}$$

$$+ \frac{(\sum X_3)^2}{N_3} - \frac{(\sum_{tot} X)^2}{N_{tot}}$$

$$= \frac{(420)^2}{10} + \frac{(530)^2}{10}$$

$$+ \frac{(640)^2}{10} - \frac{(1590)^2}{30}$$

$$= \frac{176400}{10} + \frac{280900}{10}$$

$$+ \frac{409600}{10} - 84270$$

$$= \frac{866900}{10} - 84270$$

$$= 2420$$

$$SS_{wg} = \sum_g \left[(\sum X_g{}^2) - \frac{(\sum X_g)^2}{N_g}\right]$$

$$= 20448 - \frac{(420)^2}{10} +$$

$$29884 - \frac{(530)^2}{10} +$$

$$43618 - \frac{(640)^2}{10}$$

$$= 2808 + 1794 + 2658$$

$$= 7260$$

squared deviations involved in the former. For this reason a large SS may result either from a large variability of the characteristic being measured or from a large N in the sample. Consequently, the interpretation of an SS value is more complicated than the interpretation of SD, for example. In this respect an SS is an unsatisfactory measurement of variability and can only be of limited usefulness, unless it leads to some new statistic that does not depend so greatly upon N.

Fortunately, such a statistic does exist, and we have had some previous acquaintance with it. Recall that when we first discussed SD, the *variance* was defined as equal to SD^2. However, in the present context we will call this quantity the *mean square* rather than the variance, because *mean square* (MS) has come to be used in place of variance in discussing the analysis of variance technique. As we said when we discussed the ABC's of analysis of variance, we want to judge whether our group $\overline{X}$'s differ significantly by comparing a measure of variability between groups, which we now surmise to be a mean square between groups, with a measure of variability within groups, which will be the mean square within groups. We then discover how large a value of the ratio of these two variabilities is required to indicate that the group $\overline{X}$'s differ more than would be expected if the null hypothesis of equal u's were true. With this in mind, it appears that what we want now is to find one MS (mean-square) value to represent variability between groups and another MS value to represent variability within groups. Once these values (MS_{bg} and MS_{wg} respectively) are known for any experiment, the size of their ratio can be used as the basis for drawing statistical conclusions about the data being analyzed.

To find an MS value, we simply divide the appropriate SS by its *df*, just as we divided $\sum x^2$ by N to find $\tilde{\sigma}^2$ in Chapter 5. MS_{bg} is based upon SS_{bg} and df_{bg}, the number of degrees of freedom between groups:

$$MS_{bg} = \frac{SS_{bg}}{df_{bg}} \tag{12.8}$$

where $df_{bg} = k - 1$. We note that df_{bg} has a meaning similar to that of *df* in the *t* test. However, it is based upon k, the number of groups, rather than upon the number of subjects as with the *t* test.

The second MS value, MS_{wg}, is the ratio of SS_{wg} to df_{wg}:

$$MS_{wg} = \frac{SS_{wg}}{df_{wg}} \tag{12.9}$$

where $df_{wg} = N_{tot} - k$.

We may easily remember the value of df_{wg} by saying that the *df* within a single group is one less than the number of scores in the group (that is, $N_g - 1$), thus making the *df* within all k groups equal to $(\sum N_g) - k$ or $N_{tot} - k$, since 1 *df* is lost in each of the k groups.

We now illustrate the determination of MS_{bg} and MS_{wg} with a computational example. By referring to Table 12.1 we find that $SS_{bg} = 2420$ and $SS_{wg} = 7260$ in our hypothetical attitude experiment. Since $N_{tot} = 30$ and $k = 3$ for this experiment, $df_{bg} = 3 - 1 = 2$ and $df_{wg} = 30 - 3 = 27$. First we find MS_{bg}:

$$MS_{bg} = \frac{2420}{2} = 1210$$

Then we compute MS_{wg}:

$$MS_{wg} = \frac{7260}{27} = 268.9$$

MS_{bg} will not be greatly larger than MS_{wg} in an experiment unless the group $\overline{X}$'s differ significantly. We have just found an MS_{bg} of 1210, compared with an MS_{wg} of 268.9, which leads us to suspect that the variability between group $\overline{X}$'s is too great to be attributed to random differences between samples from populations with identical means. To make a more precise comparison of these MS values we now form what is called the F ratio and evaluate it.

THE F RATIO

The F ratio, which is simply a numerical expression of the relative size of MS_{bg} and MS_{wg}, is defined by the equation below:

$$F = \frac{MS_{bg}}{MS_{wg}} \tag{12.10}$$

When an F ratio has been found in an experiment, a decision regarding the significance of the differences in group $\overline{X}$'s is made by comparing that ratio with the F values to be expected if the null hypothesis that the population μ's are equal were true. The information necessary for this comparison is presented in Table F at the end of the book. (Note the happy accident that the values of F appear in Table F.) Table F shows the F values required at the 5% and 1% levels of significance. That is, this table presents the F values that are so large that they would be exceeded only 5% or 1% of the time by random sampling if the null hypothesis were true. If MS_{bg} is enough greater than MS_{wg}, so that the resulting F ratio exceeds the tabled F at the 5% (or 1%) level, then we realize that it would be very unlikely that such an F would occur unless the null hypothesis were false. Therefore, we would reject the null hypothesis that the population μ's are equal and say that our obtained F is significant at the 5% (or 1%) level.

One step in this procedure remains to be explored: we must know where to look in Table F for the significant values of the F ratio with which to compare the F obtained in a particular experiment. This step is much like the procedure for finding a significant t value in Table C. There we had to look for a t associated with a df value. Now we must find an F associated with two df values, because the degree of significance of any obtained F depends upon two factors—the number of groups and the number of subjects within the groups. Accordingly, we employ the values of df_{bg} and df_{wg} in using Table F.

Knowing that $df_{bg} = 2$ and $df_{wg} = 27$ in our attitude experiment, we can find a cell in Table F that is appropriate for that experiment. Since the df for the MS in the numerator of the F ratio is df_{bg} and the df for the MS in the denominator is df_{wg}, we see from the table that we must go to the *column* labeled $df = 2$ and proceed down to the *row* labeled $df = 27$. There we find two F values, one for the 5% level of significance and the other for the 1% level. The F ratio at the 5% level of significance is 3.35 and at the 1% level, 5.49. An F obtained in an experiment with three groups, each of ten subjects, must be compared with these values to determine whether or not it is significant.

To illustrate the use of the F ratio, we again employ our data from the attitude experiment. Since $MS_{bg} = 1210$ and $MS_{wg} = 268.9$, we find F with the following equation:

$$F = \frac{1210}{268.9} = 4.50$$

This F ratio of 4.50 is significant at the 5% level because it is greater than 3.35, the F value tabled at the 5% level for 2 and 27 df. However, it is not significant at the 1% level because it is less than 5.49, the value required for significance at that level. Thus, in this situation we would reject the null hypothesis if we were using a 5% level of significance but accept it if we were using a 1% level. We may notice that, if df_{wg} had been between 30 and 40 rather than at 27 as it was, Table F would not have indicated the exact F values required for significance at the 5% and 1% levels. In that case we would have used the tabled values for 2 and 30 df, because they are higher than those for 2 and 40 df, thereby behaving with extra caution when we conclude that an F is significant at the 5% or 1% level.

Assumptions Underlying the Analysis of Variance

Like every other statistical test we have studied, the analysis of variance involves certain assumptions which had to be made in order to derive the table of significant values for the test. Table F is known to present the F values required for significance at the 5% and 1% levels when three assumptions are satisfied: (a) each of the k populations from which the groups in the experiment were drawn is normally distributed, (b) the σ^2 values for the k populations are equal, (c) the subjects of the experiment have been randomly and independently drawn from their respective populations.

If we know that these assumptions have been met, we can accept conclusions based upon the use of Table F at their face value. If these assumptions are not satisfied, the mathematical demonstrations that the values presented in Table F are the exact values required at the 5% and 1% levels can no longer be made. However, the practical usefulness of the analysis-of-variance procedure may be nearly as great when one or two of these assumptions are fulfilled as when all are satisfied. If one of these assumptions appears not to be met, the experimenter may prefer to perform the analysis of variance and interpret it conservatively (for example, require that his F values reach the tabled values for a higher level for significance such as 1% when he would otherwise have required only a 5% level for concluding that F is significant). He chooses to do this rather than to discard his data or to seek a new statistical procedure.

To decide whether the assumptions of normality and equal σ^2's have been met is not easy. Since the groups involved in this type of experiment are usually small, there are seldom enough scores to indicate the shape of the distribution for each population. If the sample distribution is very far from normal, we may suspect that the distribution of the population is not normal. Similarly, very great differences among the $\tilde{\sigma}^2$'s for the different groups suggest that the population σ^2's may be unequal. Statistical tests of normality and equality of σ^2's could be presented here; we omit them because of the space they would require and the impracticability of applying a test of normality except when N is large.

A Step-by-Step Computational Procedure

We now can list in sequence all of the steps required to perform a simple analysis of variance. Then we will illustrate each step by analyzing the results of an actual experiment.

Basic data

(1) Find for each separate group $\sum X$, $\sum X^2$, and N.

(2) By summing the appropriate figures for each of the groups, find $\sum_{tot} X$, $\sum_{tot} X^2$, and N_{tot}.

Calculational steps

Find the SS's by the formulas:

$$(3) \ SS_{tot} = \left(\sum_{tot} X^2 \right) - \frac{(\sum_{tot} X)^2}{N_{tot}}$$

(4) $SS_{bg} = \sum_g \left[\dfrac{(\sum X_g)^2}{N_g} \right] - \dfrac{(\sum X)^2_{tot}}{N_{tot}}$

(5) $SS_{wg} = SS_{tot} - SS_{bg}$

or

$$SS_{wg} = \sum_g \left[(\sum X_g^2) - \dfrac{(\sum X_g)^2}{N_g} \right]$$

Find *df*'s by the formulas:

(6) $df_{bg} = k - 1$

(7) $df_{wg} = N_{tot} - k$

Find MS's by the formulas:

(8) $MS_{bg} = \dfrac{SS_{bg}}{df_{bg}}$

(9) $MS_{wg} = \dfrac{SS_{wg}}{df_{wg}}$

Find and determine the significance of F:

(10) $F = \dfrac{MS_{bg}}{MS_{wg}}$

(11) Entering Table F with the df_{bg} and df_{wg} found in Steps 6 and 7, find the F's required for significance at the 5% and the 1% levels.

(12) If your obtained F exceeds the required F at the 5% level (or the 1% level) reject the null hypothesis at that level; otherwise, accept it.

We can illustrate these steps by referring to the results of an experiment[2] in which four groups of rats were *trained* under different degrees of hunger and then *tested* under a single hunger condition, 1 hour of food deprivation. Group 1 was trained under 1 hour of hunger, Group 2 under 7 hours of hunger, Group 3 under 15 hours of hunger, and Group 4 under 22 hours of hunger. Let us perform an analysis of variance to find out whether the four hunger conditions during training produced significantly different numbers of correct responses during test trials. Table 12.2 presents data for these four groups together with the X^2 value for each X.

[2] Teel, K. S. Habit strength as a function of motivation during learning. *Journal of Comparative and Physiological Psychology*, 1952, 45, 188–191. Raw data courtesy of the author.

Teel's experiment involved sixteen groups. Because the complexity of the experiment precludes its complete presentation here, we refer here to only four of his groups, treating them as if they represented a complete experiment.

Table 12.2 Preliminary steps in the analysis of variance of the number of correct test responses (X) by rats trained under four hunger conditions.

Group I 1 hr. hunger		Group II 7 hrs. hunger		Group III 15 hrs. hunger		Group IV 22 hrs. hunger	
X_1	X_1^2	X_2	X_2^2	X_3	X_3^2	X_4	X_4^2
24	576	16	256	8	64	28	784
24	576	8	64	36	1296	8	64
16	256	28	784	16	256	20	400
12	144	24	576	20	400	8	64
32	1024	48	2304	24	576	8	64
		8	64				
$\sum X_1 = 108$	$\sum X_1^2 = 2576$	$\sum X_2 = 132$	$\sum X_2^2 = 4048$	$\sum X_3 = 104$	$\sum X_3^2 = 2592$	$\sum X_4 = 72$	$\sum X_4^2 = 1376$

$$\sum_{tot} X = \sum X_1 + \sum X_2 + \sum X_3 + \sum X_4 = 108 + 132 + 104 + 72 = 416$$

$$\sum_{tot} X^2 = \sum X_1^2 + \sum X_2^2 + \sum X_3^2 + \sum X_4^2 = 2576 + 4048 + 2592 + 1376 = 10592$$

$$N_{tot} = N_1 + N_2 + N_3 + N_4 = 5 + 6 + 5 + 5 = 21$$

Basic data

Step 1 is to find $\sum X$, $\sum X^2$, and N for each group, and Step 2 to add these quantities together to get $\sum\limits_{tot} X$, $\sum\limits_{tot} X^2$, and N_{tot}. These values are presented in Table 12.2.

Calculational steps

The SS's are first calculated as follows:

(3) $SS_{tot} = \left(\sum\limits_{tot} X^2\right) - \dfrac{\left(\sum\limits_{tot} X\right)^2}{N_{tot}}$

$= 10592 - \dfrac{(416)^2}{21} = 10592 - 8240.76$

$= 2351.24$

(4) $SS_{bg} = \sum\limits_{g} \left[\dfrac{(\sum X_g)^2}{N_g}\right] - \dfrac{\left(\sum\limits_{tot} X\right)^2}{N_{tot}}$

$= \left[\dfrac{(108)^2}{5} + \dfrac{(132)^2}{6} + \dfrac{(104)^2}{5} + \dfrac{(72)^2}{5}\right] - \dfrac{(416)^2}{21}$

$= 8436.80 - 8240.76$

$= 196.04$

(5) $SS_{wg} = SS_{tot} - SS_{bg}$

$= 2351.24 - 196.04$

$= 2155.20$

Then df's and MS's are found as follows:

(6) $df_{bg} = k - 1 = 4 - 1 = 3$

(7) $df_{wg} = N_{tot} - k = 21 - 4 = 17$

(8) $MS_{bg} = \dfrac{SS_{bg}}{df_{bg}} = \dfrac{196.04}{3}$

$= 63.35$

(9) $MS_{wg} = \dfrac{SS_{wg}}{df_{wg}} = \dfrac{2155.20}{17}$

$= 126.78$

Finally, the F is obtained and its significance determined:

(10) $F = \dfrac{MS_{bg}}{MS_{wg}} = \dfrac{65.35}{126.78} = .52$

(11) Looking in Table F for the case where $df_{bg} = 3$ and $df_{wg} = 17$, we find that F must be 3.20 or greater to be significant at the 5% level, and 5.18 or greater to be significant at the 1% level.

(12) Our F of .52 is not significant, since it is much less than the F of 3.20 required for significance at the 5% level. In fact, as Table F shows, no F below 1 is significant regardless of the df values. Therefore, we cannot reject the null hypothesis that the groups come from populations with the same μ.

Table 12.3 Summary of the analysis of variance of the experiment on effects of food deprivation in animals.

Source of variation	SS	df	MS	F	p
Between groups	196.04	3	65.35	.52	> .05
Within groups	2155.20	17	126.78		
Total	2351.24	20			

Table 12.3 summarizes the results of our analysis of variance, listing the SS's, df's, MS's, and the F that we have calculated as well as the p value of F. This table is an example of the standard method used to summarize an analysis of variance.

A MULTIPLE COMPARISON TEST FOLLOWING ANALYSIS OF VARIANCE

When we have performed an analysis of variance and found a significant F between the groups, we are able to reject the hypothesis that all of the groups were drawn from populations with the same mean. But, when more than two groups are being compared, a significant F does not imply that each sample $\overline{X}$ necessarily differs significantly from every other sample $\overline{X}$; quite often we are interested in pinpointing more precisely where the differences lie. Our experimental interests, for example, may lead us to inquire about the significance of the differences between specific pairs of means. Or in other instances we may observe that the data pattern themselves in interesting ways. We may find, for example, that the means appear to fall into subsets or clusters, and we may want to determine whether the clusters differ significantly from each other, or we may find that one of the means is markedly different from all the rest. What is required, then, is some technique that will allow us to follow up the analysis of variance with further statistical comparisons.

Although t might seem an appropriate test for this purpose, several statistical difficulties prohibit its use, as we noted earlier. But a number of other tests for making multiple comparisons following analysis of variance have been suggested by statisticians. Here we present one such multiple-comparison test, a method devised by Tukey,[3] which we will call the *honestly significant difference* test or *hsd*.

The Tukey *hsd* usually is performed only if the F obtained in the analysis of variance is significant, but it is theoretically permissible to perform whatever the significance of F. The test has the advantage of allowing us to compare any pair of means, or any pair of subsets of means, that we wish to select and of placing no limits on the number of comparisons that may be made with any set of data. We will illustrate its use only for the simple case in which we wish to compare individual pairs of means and in which the N's in each group are equal. Procedures are available that permit comparisons of subsets of means (e.g., $\bar{X}$'s of Groups 1 and 2 vs. $\bar{X}$'s of Groups 3, 4, and 5) and of groups of unequal size, but they are more complex.

The test involves computing a quantity that we will identify as *hsd*. The difference between any pair of means is significant at a given α level if it equals or exceeds the absolute value of *hsd* we have computed.

The formula for *hsd* is as follows:

$$hsd = q_\alpha \sqrt{\frac{MS_{wg}}{N_g}} \tag{12.11}$$

In this equation, MS_{wg} is the within-groups mean square calculated in the analysis of variance, and N_g is the number of cases in each group. The value of q_α, also called the Studentized range statistic, is found from the appropriate entry in Table I at the back of the book. We enter Table I at the specified α level for the *df* associated with MS_{wg} and k, the total number of groups in the experiment. Suppose, for example, that we wished to compare one or more pairs of means, using $\alpha = .05$, from a 4-group experiment in which there were 5 subjects per group. Looking at Table I for $k = 4$ and $df_{wg} = 16$, we find that $q_{.05} = 4.05$.

We will illustrate the application of the test by using the data from the hypothetical attitude experiment reported in Table 12.1 that involved 3 groups of 10 cases each. We repeat below some of the results obtained in our analysis of these data that we will need to determine *hsd*.

$\bar{X}_1 = 42$ $\qquad$ $\bar{X}_2 = 53$ $\qquad$ $\bar{X}_3 = 64$

$MS_{wg} = 268.9$ $\qquad$ $df_{wg} = 27$ $\qquad$ $k = 3$

where $k =$ the total number of groups and $df_{wg} = N_{tot} - k$.

[3] Tukey, J. W. The problem of multiple comparisons, mimeographed, Princeton, N.J., 1953. The computational procedures are adapted from Runyon, R. P., and Haber, A. *Fundamentals of Behavioral Statistics*, Second Edition, Addison-Wesley, 1971.

The first step (assuming we wish to compare all possible pairs of means) is to find the *absolute difference* between each pair of $\overline{X}$'s (e.g., the absolute difference between $\overline{X}_1$ and $\overline{X}_2$ is $53 - 42 = 11$). In Table 12.4 we show these differences for our experiment.

Table 12.4   Absolute differences between pairs of means in hypothetical attitude experiment.

	$\overline{x}_1$	$\overline{x}_2$
$\overline{X}_2$	11	——
$\overline{X}_3$	22	11

The next step is to compute *hsd*. Suppose that we set $\alpha = .05$. In order to find $q_{.05}$ we consult Table I to find the value associated with this alpha level for $df_{wg} = 27$ and $k = 3$. Unfortunately, by looking down the *df* column at the left, we discover that just as in the case of the F table, Table I is incomplete and has no entry for $df = 27$. Again, we choose to be conservative and use the next lowest *df*, which turns out to be 24. Going across this row to the $k = 3$ column, we find that the tabled value for *q* at $\alpha = .05$ is 3.53. We now enter this value, along with the values of MS_{wg} and N_g, into the *hsd* equation.

$$hsd = q_{.05} \sqrt{\frac{MS_{wg}}{N_g}} = 3.53 \sqrt{\frac{268.9}{10}} = 3.53\,(5.19) = 18.32$$

We have now computed the critical value that the difference between a pair of means must equal or exceed in order to be significant at the 5% level. By inspecting Table 12.4 we see that the difference between $\overline{X}_1$ and $\overline{X}_3$ is the only one that exceeds 18.32. The mean differences for the remaining pairs are not significant.

We said above that the Tukey test places no limits on the number of comparisons that may be made with any set of data. By this we mean that the probability of finding a significant difference by chance at a given α level does not increase with each additional comparison, as it does in the case of *t*. When all population means are equal and a series of Tukey multiple-comparison tests such as the three *hsd* tests just reported are made at a given α level (let us say, .05), the probability is .05 that a Type I error will occur *anywhere* in the set of comparisons that are to be performed. Since the 5% significance level applies to the *total* experiment, and not just to individual comparisons, we can perform as many comparisons as we choose and remain confident that we are not increasing our chances of a Type I error.

In concluding this chapter, we should mention that we have presented only the most elementary form of analysis of variance. This simple analysis of variance, as it is called, is the type used for testing the null hypothesis that the population means are equal in experiments involving several groups but only one basic variable, for example, the number of courses in music appreciation, number of hours of food deprivation, methods of teaching French, and so on. In the next chapter we will discuss an extension of the simple analysis of variance technique, one that is used to analyze the data from a more complicated type of experiment in which the effects of not one but two variables are simultaneously studied.

TERMS AND SYMBOLS TO REVIEW

Simple analysis of variance

Total variability

Variability within groups

Variability between groups

Sum of squares within groups (SS_{wg})

Sum of squares between groups (SS_{bg})

Sum of squares for total (SS_{tot})

Mean Square (MS)

F ratio

Assumptions

Tukey *hsd* multiple comparison test

Double-Classification Analysis of Variance

13

All the types of experiments we have considered so far have involved two or more groups differing with respect to some single variable or factor. In a good many experiments, however, the investigator studies simultaneously the effects of more than one variable. For example, in a study of animal learning such as we considered in the last chapter, we might have groups that are trained under 1, 7, 15, or 22 hours of food deprivation just as before but divide each of these deprivation groups into two subgroups, one being given a large food reward on each training trial and the other a small one. This general arrangement, in which we study the effects of two or more variables, is known as a *factorial design*. In this chapter we will be considering the analysis of variance appropriate for data from experiments employing a two-way or *double-classification* factorial design—that is, experiments in which two variables are simultaneously manipulated.

We have seen that experiments investigating a single variable need not compare only two groups but may involve any number. Similarly, there is no limit to the number of groups representing each variable in a double-classification factorial study. In our animal learning experiment, for example, we suggested studying four degrees of hunger and two magnitudes of food reward, a design calling for 4 × 2 or 8 independent groups. We could, however, extend our study to include more conditions within each variable, such as 0, 1, 2, 7, 15, and 22 hours of deprivation, and small, medium, and large reward, thus calling for 6 × 3 or 18 groups. As we increase the number of specific conditions within each type of variable, the total number of groups that needs to be tested to form a complete factorial design increases quite drastically, requiring considerable time and effort to conduct the experiment. However, the logic of the double-classification design and the statistical methods we use to analyze the experimental results remain the same, whatever the number of experimental conditions within each of the variables.

A word now about terminology. We will refer to the two kinds of conditions being manipulated as *variables*, using the letters A and B as identifying labels. The specific experimental conditions employed in investigating each of the variables we will identify by the numerals I, II, III, and so on to the nth condition. Many of the variables investigators wish to study involve some kind of quantifiable dimension such as time, distance, or amount, so that the specific conditions being investigated can be rank-ordered in magnitude from least to most, and the numerals I, II, III, . . . , n used to reflect this order. Often, however, the variables that we study involve two or more discrete categories or types of conditions. We may, for example, be interested in comparing several different kinds of teaching methods, persons of different nationalities, men versus women, and so on. In these instances the groups or conditions obviously cannot be rank-ordered along some dimension, so that our use of identifying numerical labels is purely nominative, the assignment of number to group being completely arbitrary.

MAIN EFFECTS AND INTERACTION IN ANALYSIS OF VARIANCE

Application of the analysis of variance technique to the data from a double-classification factorial experiment allows us to make three kinds of statements about our results: (1) the effects on our response measure of the different conditions of Variable A, independent of variations in B conditions (*main effects of A*), (2) the effects of the different conditions of Variable B, independent of variations in A conditions (*main effects of B*), and (3) the joint effects or *interaction* of Variables A and B. Rather than proceeding directly to the statistical steps that will allow us to make these comparisons, we first will explain what each of them involves.

Table 13.1 presents the results of three hypothetical experiments employing 2 × 2, 2 × 3, and 3 × 3 factorial designs, respectively, with equal N's per group in each experiment. *(Although these are methods for dealing with data from experiments employing unequal numbers of subjects*

Table 13.1 Two-way tables showing means for hypothetical 2 × 2, 2 × 3, and 3 × 3 factorial experiments, with equal N's per group.

N = 5
Variable A

		I	II	$\bar{X}_{Row}$
Variable	I	20	27	23.5
B	II	26	18	22
$\bar{X}_{Col}$		23	22.5	

N = 10
Variable A

		I	II	III	$\bar{X}_{Row}$
Variable	I	3	7	8	6
B	II	5	9	10	8
$\bar{X}_{Col}$		4	8	9	

N = 5
Variable A

		I	II	III	$\bar{X}_{Row}$
Variable	I	29	30	31	30
B	II	30	34	36	33.33
	III	31	38	42	37
$\bar{X}_{Col}$		30	34	36.33	

in each group, we will only consider techniques that are appropriate for equal or nearly equal N's.) The numbers entered in the cells are the means obtained for each of the specific groups. For each table we have also reported the mean of all the cases represented in each column and in each row. For example, in the 3 × 3 experiment whose results are reported in Table 13.1, we added together the scores of all individuals tested under the AI condition, whatever their B condition, obtaining a sum of 450. This $\sum X$ was then divided by 15, the total number of AI subjects, yielding a column mean of 30. (Since the N's in each cell are equal, we could also get the column $\overline{X}$ by averaging the three cell $\overline{X}$'s.) The means of all the AII and of all the AIII cases, shown at the bottom of the second and third columns, were obtained in a similar fashion, as were the three row means, shown at the right of the table. These row values, of course, are the means of all 15 of the cases tested under the BI, BII, and BIII conditions, whatever their A condition.

Main Effects

Determination of the main effects of Variable A and Variable B by means of a double-classification analysis of variance involves testing a particular null hypothesis about the column $\overline{X}$'s and the row $\overline{X}$'s. This hypothesis states that the means of the populations from which the samples were drawn are all equal ($\mu_1 = \mu_2 = \cdots = \mu_N$). Thus H_0 specifies that each A group, averaged over all B groups, is drawn from a population with the same mean and that therefore any differences among the $\overline{X}$'s of the A groups (column $\overline{X}$'s) are due to chance factors. Similarly, any differences among the $\overline{X}$'s of the B groups (row $\overline{X}$'s) are also due to chance.

A method of testing these null hypotheses may occur to you immediately. In comparing the means of the A groups (column $\overline{X}$'s), why not ignore the B variable and consider all those tested under the same A condition (all those in the same column) as forming a single group? Then we could perform a simple analysis of variance of the A groups, following the procedures outlined in Chapter 12. Another simple analysis of variance could then be performed in the same way on the row data, allowing us to test the null hypothesis about the means of the B groups. In conducting this type of analysis, you will recall, we find an F ratio that compares the relative size of two measures of variability: the variability between the means of the groups and the variability within each of the groups, the latter being due to uncontrolled, unidentified factors such as original differences in ability among our subjects. If the between-groups measure is sufficiently larger than the within-groups measure, we reject the null hypothesis.

However, if we were to perform these simple analyses of variance on the data from a double-classification experiment, we would run into a major problem. Our measure of within-groups variability for the A (column) variable would reflect differences among the subjects in each column that are due not only to uncontrolled, unidentified factors but also to any systematic effect of variations in B (row) conditions. In like manner, our within-groups measure for the row data would reflect any differences among subjects due to the effects of variations in A conditions (columns). These systematic effects would inflate each of the measures of within-group variability and hence decrease the size of the F ratio. To avoid this complication, we modify our simple analysis-of-variance approach by computing a *single* within-groups measure based on each of our separate experimental groups—that is, within each *cell* of our two-way table. We then obtain our F ratios for the main effects of Variables A and B by dividing each of our measures of between-groups variability (one between column $\overline{X}$'s, the other between row $\overline{X}$'s) by this single measure of within-groups (within-cells) variability.

Interaction Between Variables

The reason we conduct a factorial study is not simply to investigate the main effects of more than one variable. One of the major advantages of the factorial experiment is that it allows us to study the *joint effects* of the two variables, how the variables combine or *interact* to influence our response measure.

The ways in which variables can combine are numerous. One simple type of combination is an *additive* one—a combination of the form X = A + B. The data from the hypothetical 2 × 3 experiment reported in Table 13.1 illustrate this type of combination. We can most easily explain an additive relationship by referring to the graphic presentation of these data in Figure 13.1. We have shown the three A conditions along the baseline and have plotted the means of the three BI groups and the three BII groups in separate curves. As you can see by inspection, the curves for the BI and BII groups are *parallel.* This fact indicates that the magnitude of the effect

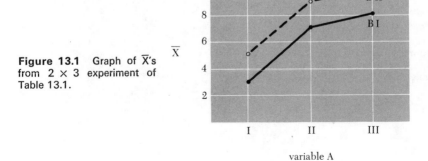

Figure 13.1 Graph of $\overline{X}$'s from 2 × 3 experiment of Table 13.1.

of the B variable on our response measure is the *same* for each condition of A. An alternate way of describing this type of relationship is to state that the difference in X̄'s between BI and BII groups tested under the same A condition is the same for each of the A conditions. In this example, as you can easily see from the values reported in Table 13.1, the difference in X̄'s (BII − BI) at each level of A is 2 (5 − 3 = 2, 9 − 7 = 2, 10 − 8 = 2). We could have graphed the data the other way around, representing the two B conditions along the baseline and plotting the means of the AI, AII, and AIII groups in separate curves. As you may determine for yourself by graphing the data in this manner, the three curves would again turn out to be parallel. You can also see that this would be so by examining the means of groups shown in Table 13.1. The difference between the X̄'s of any given pair of A groups (I versus II, II versus III, or I versus III) is the same whatever the B condition under which the pair was tested. For AI versus AII, for example, the difference at both levels of B is 4.

The data from the 2 × 2 and 3 × 3 experiments reported in Table 13.1 are examples of two kinds of *nonadditive* combinations, as can again be conveniently demonstrated by presenting their results graphically. Looking first at the data of the 3 × 3 experiment plotted in Figure 13.2, you will observe that the curves for the three A groups are not parallel; rather, they diverge. That is, the magnitude of the differences between the A groups becomes larger as one goes from BI to BII and from BII to BIII. Had the curves for the three B groups been plotted instead, these curves would also have been shown to diverge with changes in A condition.

An example of a very complex type of nonadditive interaction is found in the data of the 2 × 2 experiment reported in Table 13.1. Scores on our response measure are affected by both A and B conditions but, as you can see from the plot of the data in Figure 13.3, the *direction* of the effect of variations in A condition is different for each B condition (and vice versa). Strange as it may seem at first glance, this type of interaction occurs not

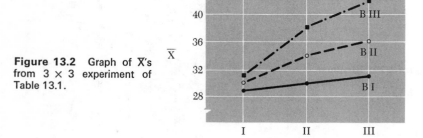

Figure 13.2 Graph of X̄'s from 3 × 3 experiment of Table 13.1.

variable A

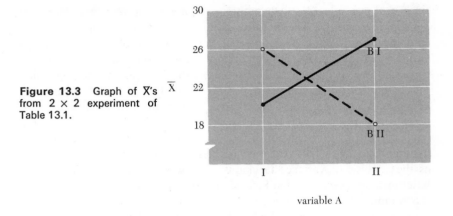

Figure 13.3 Graph of $\overline{X}$'s from 2 × 2 experiment of Table 13.1.

variable A

infrequently in psychological data. For example, individuals who are given a long list of very simple arithmetic problems to solve with instructions that emphasize the importance of speed may produce not only more answers in a given period of time but more *correct* answers than individuals given the same problems with instructions that emphasize the importance of accuracy more than speed. The same sets of instructions, however, may produce the *opposite* results with respect to correct responses when a list of complicated problems is used; individuals given speed instructions might make all sorts of errors in their attempt to hurry and therefore arrive at a smaller number of correct solutions than the group trying to be accurate. Thus, speed instructions may lead to either better or worse performance than accuracy instructions, depending on the type of problems employed; the direction of the effects of variations in A is dependent on the particular B condition under which subjects were tested.

One by-product of this type of complex interaction should be mentioned. Since the direction of the influence of one variable is dependent on the other variable, comparisons of the overall means of the various A groups, ignoring B conditions, or of the overall means of the B groups, ignoring A conditions, may reveal that neither variable has any significant main effect. You can see, for example, that in the 2 × 2 experiment shown in Table 13.1 the two column means are very similar in value, as are the two row means. We would not conclude, however, that neither variable has any kind of effect but rather (assuming that the relevant statistical term is significant) that the *direction* of the effects of one variable is dependent on the specific condition employed with respect to the second variable.

A major purpose in conducting a double-classification factorial study, as we have said, is to determine how the two variables combine. The double-classification analysis-of-variance technique is of assistance in this process since it allows us to test a hypothesis about the nature of the interaction between the variables. The specific hypothesis being tested is that the variables combine in an *additive* fashion—that the magnitude of the effects of one variable is constant over all conditions of the second variable. Expressing this in another way, we test the hypothesis that the curves resulting from a plot of the $\bar{X}$'s of the groups are *parallel*. If the F ratio found for the interaction between Variables A and B is nonsignificant at a given α level, we *accept* the hypothesis that the variables combine additively. If, on the other hand, the F for the interaction between variables is significant, we *reject* this hypothesis and conclude that the variables combine nonadditively. Versatile as the analysis of variance technique is, however, it cannot tell us in this latter instance precisely what kind of nonadditive relationship has occurred; this we must determine by inspection of our data.

In summary, then, a double-classification analysis of variance yields F ratios that allow us to test the null hypothesis that μ's are equal concerning differences between the means of the A groups and between the means of the B groups—the *main effects* of each of the two variables—and a hypothesis concerning the *interaction* between the two variables: that they combine in an additive manner. We turn now to the computational steps involved in conducting such an analysis. Remember that we are presenting formulas for the case in which all groups have the same N.

COMPUTATIONAL STEPS IN DOUBLE-CLASSIFICATION ANALYSIS OF VARIANCE

In describing the procedures employed in performing a double-classification analysis of variance we will omit the basic formulas defining the various sums of squares and present only the computational formulas, illustrating their application with the data from the hypothetical experiment shown in Table 13.2. In this experiment, the minimum exposure time (duration threshold) required for correct identification of words was being investigated as a function of several word characteristics. Three groups of ten individuals were given lists of words that had been judged to be high, medium, or low in familiarity; half the subjects in each of these familiarity groups were given a list containing words that were emotionally unpleasant in meaning and the other half a list of emotionally neutral words. The experiment thus formed a 2×3 factorial design with five subjects in each of the six groups. The score obtained for each of the subjects was the sum of minimum exposure times for the various words on the list.

In Table 13.2 we have also reported the $\sum X$ and $\sum X^2$ for each of the six groups in our hypothetical experiment, since these are the basic data

Table 13.2 Data from the hypothetical study of perceptual thresholds.

Neutral						Unpleasant					
High		*Medium*		*Low*		*High*		*Medium*		*Low*	
X	X²	X	X²	X	X²	X	X²	X	X²	X	X²
13	169	21	441	24	576	16	256	22	484	32	1024
19	361	14	196	15	225	14	196	19	361	27	729
21	441	16	256	22	484	21	441	16	256	25	625
16	256	21	441	19	361	19	361	20	400	27	729
11	121	13	169	17	289	23	529	29	841	21	441
80	1348	85	1503	97	1935	93	1783	106	2342	132	3548

$$\sum_{tot} X = 80 + 85 + 97 + 93 + 106 + 132 = 593$$

$$\sum_{tot} X^2 = 1348 + 1503 + 1935 + 1783 + 2342 + 3548 = 12459$$

Summary of $\sum X$ values

Word familiarity (Variable A)

		I (high)	II (med.)	III (low)	sum
Affectivity	I (neutral)	80	85	97	262
(Variable B)	II (unpleasant)	93	106	132	331
	Sum	173	191	229	593

entering into our calculations. We will find it convenient to have the $\sum X$ values in each of the groups set up in a two-way table so that we can determine the sums for each column and row, as well as $\sum_{tot} X$. This type of two-way table is also shown in Table 13.2.

Computational Steps in Finding SS's

You will remember that in performing a simple analysis of variance on the data from a collection of k groups, we first find the total sum of squares, the SS for all the subjects in the experiment. We then proceed to break down SS_{tot} into its two components: SS_{bg} and SS_{wg}. Our first step in analyzing the results of a double-classification factorial experiment is to treat the data as a collection of k groups and find SS_{tot}, SS_{bg}, and SS_{wg} using exactly the same procedures as we employed in the last chapter. Illustrating these procedures[1]

[1] Note that formulas (13.1–13.4) are the same as (12.4–12.7) on pp. 185–186.

with the data from our 2×3 experiment in Table 13.2 (where $k = 6$) we thus have:

$$SS_{tot} = \left(\sum_{tot} X^2 \right) - \frac{(\sum_{tot} X)^2}{N_{tot}} \tag{13.1}$$

$$= 12459 - \frac{(593)^2}{30} = 12459 - 11721.63$$

$$= 737.37$$

$$SS_{bg} = \sum_{g} \left[\frac{(\sum X_g)^2}{N_g} \right] - \frac{(\sum_{tot} X)^2}{N_{tot}} \tag{13.2}$$

$$= \left[\frac{(80)^2}{5} + \frac{(85)^2}{5} + \frac{(97)^2}{5} + \frac{(93)^2}{5} + \frac{(106)^2}{5} + \frac{(132)^2}{5} \right] - \frac{(593)^2}{30}$$

$$= 12068.60 - 11721.63 = 346.97$$

$$SS_{wg} = SS_{tot} - SS_{bg} \tag{13.3}$$

$$= 737.37 - 346.97 = 390.40$$

We will remind you that we could also have computed SS_{wg} directly rather than by subtraction, using the formula:

$$SS_{wg} = \sum_{g} \left[(\sum X_g^2) - \frac{(\sum X_g)^2}{N_g} \right] \tag{13.4}$$

The SS_{wg} we have just determined is based on the variability within each of our six experimental groups—within each *cell* of our two-way table. As we discussed earlier, we will use this SS_{wg} to obtain a measure of within-groups variability against which to assess the main effects of both Variable A and Variable B. This same measure is also used in determining the F for the interaction of the variables.

The SS_{bg} that we have computed, however, is not of direct interest to us, since it is based on the variability among the means of all of our independent groups, considered as a single collection of k groups. This variability among the k means comes from three sources: differences in the $\overline{X}$'s of the A groups (column means), differences in $\overline{X}$'s in the B groups (row means), and the interaction between Variables A and B (that is, any departure from an additive combination of the variables). Since it is these three sources of variability that concern us, we now proceed to break down SS_{bg} and find the SS's for its three components.

The SS for Variable A is determined by the formula:

$$SS_A = \frac{(\sum X_{AI})^2}{N_{AI}} + \frac{(\sum X_{AII})^2}{N_{AII}} + \cdots + \frac{(\sum X_{Am})^2}{N_{Am}} - \frac{(\sum_{tot} X)^2}{N_{tot}} \tag{13.5}$$

This formula indicates that we first find the sum of the scores of all the individuals tested under the AI condition (the $\sum X$ of the AI column), square this sum, and divide the result by the total number of AI cases. We then find the same quantity for each of the other A groups, on to the last or $m$thA group. We then add these quantities together and subtract the value $(\sum_{tot} X)^2/N_{tot}$. (The latter quantity also appears, be sure to notice, in the formulas for SS_{tot} and SS_{bg}.)

In our illustrative experiment there were three A conditions (three degrees of word familiarity), each based on 5 BI cases and 5 BII cases for a total of 10. Thus:

$$SS_A = \frac{(\sum X_{AI})^2}{N_{AI}} + \frac{(\sum X_{AII})^2}{N_{AII}} + \frac{(\sum X_{AIII})^2}{N_{AIII}} - \frac{(\sum_{tot} X)^2}{N_{tot}}$$

$$= \frac{(173)^2}{10} + \frac{(191)^2}{10} + \frac{(229)^2}{10} - \frac{(593)^2}{30}$$

$$= 11855.10 - 11721.63$$

$$= 163.47$$

The SS for Variable B is determined in a parallel manner:

$$SS_B = \frac{(\sum X_{BI})^2}{N_{BI}} + \frac{(\sum X_{BII})^2}{N_{BII}} + \cdots + \frac{(\sum X_{Bn})^2}{N_{Bn}} - \frac{(\sum_{tot} X)^2}{N_{tot}} \qquad (13.6)$$

For the two affectivity (B) groups of our illustrative experiment, each of which is composed of $5 + 5 + 5$ or 15 cases, we thus have:

$$SS_B = \frac{(\sum X_{BI})^2}{N_{BI}} + \frac{(\sum X_{BII})^2}{N_{BII}} - \frac{(\sum_{tot} X)^2}{N_{tot}}$$

$$= \frac{(262)^2}{15} + \frac{(331)^2}{15} - \frac{(593)^2}{30}$$

$$= 11880.33 - 11721.63$$

$$= 158.70$$

You will recall that SS_{bg} can be divided into three components, representing the SS's for Variable A, Variable B, and the A × B interaction. Since we have already computed SS_{bg}, SS_A, and SS_B, we can obtain the SS for interaction by subtraction:

$$SS_{A \times B} = SS_{bg} - SS_A - SS_B \qquad (13.7)$$

For our experiment:

$$SS_{A \times B} = 346.97 - 163.47 - 158.70$$
$$= 24.80$$

Although SS for interaction is typically obtained by subtraction, it may be computed directly by the following basic formula:

$$SS_{A \times B} = N \left[\sum_g (\bar{X}_{AB} - \bar{X}_A - \bar{X}_B + \bar{X}_{tot})^2 \right] \tag{13.8}$$

where N = number of cases per group, $\bar{X}_{AB}$ = the mean of a particular group, $\bar{X}_A$ = the mean of all the cases tested under the same A condition as this group (column $\bar{X}$), $\bar{X}_B$ = the mean of all cases tested under the same B condition as this group (row $\bar{X}$), and $\bar{X}_{tot}$ = the mean of all the cases. We will leave the application of this formula to the data of Table 13.2 as an exercise for those curious enough to find out that it does indeed yield the same value as we obtained by subtraction. Remember that the hypothesis being tested concerning the A $\times$ B interaction is that the two variables combine additively. If the variables combined in perfect additive fashion, we would expect no variability due to interaction and hence that $SS_{A \times B}$ would be equal to 0. This implies in turn that the term $(\bar{X}_{AB} - \bar{X}_A - \bar{X}_B + \bar{X}_{tot})^2$ in our formula for $SS_{A \times B}$ would be 0 for each AB group. We may demonstrate this by applying the formula to the data from the 2 $\times$ 3 experiment of Table 13.1, which, as we saw earlier, were an example of a perfect additive relationship between variables. The grand mean for all of them is 7. Applying our formula to these data gives us:

$$\begin{aligned}
SS_{A \times B} &= N \left[\sum_g (\bar{X}_{AB} - \bar{X}_A - \bar{X}_B + \bar{X}_{tot})^2 \right] \\
&= 10[(3 - 4 - 6 + 7)^2 + (7 - 8 - 6 + 7)^2 + \\
&\quad (8 - 9 - 6 + 7)^2 + (5 - 4 - 8 + 7)^2 + \\
&\quad (9 - 8 - 8 + 7)^2 + (10 - 9 - 8 + 7)^2] \\
&= 0
\end{aligned}$$

Computation of MS's and df's

Having determined the SS within groups and the SS's between A groups, between B groups, and the A $\times$ B interaction (the three components making up SS_{bg}), we now can find the mean square (MS) for each of these sources of variability. Just as we stated in the last chapter, these MS's are found by dividing each SS by its *df*. Thus:

$$MS_A = \frac{SS_A}{df_A} \tag{13.9}$$

where $df_A = m - 1$ (the number of A groups minus one);

$$MS_B = \frac{SS_B}{df_B}$$

where $df_B = n - 1$ (the number of B groups minus one);

$$MS_{A \times B} = \frac{SS_{A \times B}}{df_{A \times B}}$$

where $df_{A \times B} = (m - 1)(n - 1)$ or $(df_A)(df_B)$; and

$$MS_{wg} = \frac{SS_{wg}}{df_{wg}}$$

where $df_{wg} = N_{tot} - (m)(n)$ [the total number of cases minus the total number of groups—that is, A groups × B groups].

We should also note that the df for SS_{tot} is equal to $N_{tot} - 1$. Since SS_{tot} is composed of SS_{wg}, SS_A, SS_B, and $SS_{A \times B}$, the df's for these four components, when summed, equal the total df or $N_{tot} - 1$.

Applying these formulas to the data of our illustrative experiment shown in Table 13.2, we find the following df's:

$$df_A = m - 1 = 3 - 1 = 2$$

$$df_B = n - 1 = 2 - 1 = 1$$

$$df_{A \times B} = (m - 1)(n - 1) = (2)(1) = 2$$

$$df_{wg} = N_{tot} - (m)(n) = 30 - (3)(2) = 24$$

The df_{tot} in our experiment is equal to $30 - 1$ or 29, and you will observe that the four df's obtained above do indeed add up to this number.

The MS's are thus:

$$MS_A = \frac{SS_A}{df_A} = \frac{163.47}{2} = 81.74$$

$$MS_B = \frac{SS_B}{df_B} = \frac{158.70}{1} = 158.70$$

$$MS_{A \times B} = \frac{SS_{A \times B}}{df_{A \times B}} = \frac{24.80}{2} = 12.40$$

$$MS_{wg} = \frac{SS_{wg}}{df_{wg}} = \frac{390.40}{24} = 16.27$$

Computation of F Ratios

Finally we compute F ratios by dividing the MS's for A, B, and for the A × B interaction by MS_{wg}. To illustrate:

$$F_A = \frac{MS_A}{MS_{wg}} = \frac{81.74}{16.27} = 5.02 \tag{13.10}$$

$$F_B = \frac{MS_B}{MS_{wg}} = \frac{158.70}{16.27} = 9.75$$

$$F_{A \times B} = \frac{MS_{A \times B}}{MS_{wg}} = \frac{12.40}{16.27} = .76$$

The results of our calculations of the various values of SS, *df*, MS, and F are summarized in Table 13.3. Now all that remains is to evaluate each F ratio by comparing it with the value of F at the 5% or the 1% level of significance associated with the appropriate *df*'s, as listed in Table F. Looking at this table, we find that with 2 and 24 *df* (the *df*'s associated with the F for Variable A), a value of 3.40 is required at the 5% level and of 5.61 at the 1% level. Since we obtained an F of 5.02 for the A (word familiarity) groups, we conclude that these groups differed significantly at the 5% level. For 1 and 24 *df* (the *df*'s associated with the F for Variable B) we find an F of 7.82 required at the 1% level. Since our obtained F was 9.76, we conclude that the B (affectivity) groups also differed significantly, beyond the 1% level. However, the F for the A × B interaction was less than one, so obviously it is not significant. Thus, we conclude that while both word familiarity and affectivity were significantly related to duration threshold, the magnitude of the effects of one variable was the same whatever the specific condition of the other variable. The two variables, in other words, combine additively. The *p* value of each F (in terms of whether it reaches the 1% level or the 5% level, or fails to reach the latter) is shown in the final column of Table 13.3.

Table 13.3 Summary of the analysis of variance of experiment on perceptual thresholds as related to word familiarity and affectivity.

Source of variation	SS	df	MS	F	p
Familiarity (A)	163.47	2	81.74	5.02	< .05
Affectivity (B)	158.70	1	158.70	9.75	< .01
Interaction (A × B)	24.80	2	12.40	.76	> .05
Within groups	390.40	24	16.27		
Total	737.37	29			

A Step-by-Step Computational Procedure

We will now summarize the steps taken in computing a double-classification analysis of variance, starting first with a description of the basic data that will enter into our calculations.

Basic data

(1) Find the $\sum X$ of each group and enter them in a 2-way table such as is illustrated in the lower portion of Table 13.2. By adding the appropriate sums, find the $\sum X$ for each column, each row, and for the total experiment.

(2) Find the $\sum X^2$ for each group and by adding these values, for the total experiment.

Calculate SS's by the following formulas:

(3) $SS_{tot} = \left(\sum_{tot} X^2 \right) - \dfrac{(\sum_{tot} X)^2}{N_{tot}}$

(4) $SS_{bg} = \sum_{g} \left[\dfrac{(\sum X_g)^2}{N_g} \right] - \dfrac{(\sum_{tot} X)^2}{N_{tot}}$

(5) $SS_{wg} = SS_{tot} - SS_{bg}$

Check: $SS_{wg} = \sum_{g} \left[(\sum X_g^2) - \dfrac{(\sum X_g)^2}{N_g} \right]$

(6) $SS_A = \dfrac{(\sum X_{AI})^2}{N_{AI}} + \dfrac{(\sum X_{AII})^2}{N_{AII}} + \cdots + \dfrac{(\sum X_{Am})^2}{N_{Am}} - \dfrac{(\sum_{tot} X)^2}{N_{tot}}$

(7) $SS_B = \dfrac{(\sum X_{BI})^2}{N_{BI}} + \dfrac{(\sum X_{BII})^2}{N_{BII}} + \cdots + \dfrac{(\sum X_{Bn})^2}{N_{Bn}} - \dfrac{(\sum_{tot} X)^2}{N_{tot}}$

(8) $SS_{A \times B} = SS_{bg} - SS_A - SS_B$

Check: $SS_{A \times B} = N \left[\sum_{g} (\bar{X}_{AB} - \bar{X}_A - \bar{X}_B + \bar{X}_{tot})^2 \right]$

Determine *df*'s by the formulas:

(9) $df_A = m - 1$

(10) $df_B = n - 1$

(11) $df_{A \times B} = (m - 1)(n - 1)$

(12) $df_{wg} = N_{tot} - (m)(n)$

Calculate MS's by dividing each SS by its *df*:

(13) $MS_A = \dfrac{SS_A}{df_A}$

(14) $MS_B = \dfrac{SS_B}{df_B}$

(15) $MS_{A \times B} = \dfrac{SS_{A \times B}}{df_{A \times B}}$

(16) $MS_{wg} = \dfrac{SS_{wg}}{df_{wg}}$

Calculate F ratios by dividing the MS's for A, B, and A $\times$ B by MS_{wg}:

(17) $F_A = \dfrac{MS_A}{MS_{wg}}$

(18) $F_B = \dfrac{MS_B}{MS_{wg}}$

(19) $F_{A \times B} = \dfrac{MS_{A \times B}}{MS_{wg}}$

Determine the significance of F's:

(20) For each of the F ratios found in steps 17–19 find the values required for significance at the 5% and 1% level by entering Table F with the *df*'s associated with the numerator and the denominator of each of these F's.

(21) If a given F exceeds the required F at either the 5% or 1% level, reject the null hypothesis at that level; if it falls short, accept the null hypothesis.

Summarize the results:

(22) Prepare a summary table, similar to the one in Table 13.3, that contains the values of the SS, *df*, MS, and F for the main effects of variables A and B and for the A $\times$ B interaction.

If one or more of the F's is significant, we may want to make further comparisons to determine precisely which groups or conditions contributed to the significant differences. This may be done by comparing pairs of $\overline{X}$'s by the Tukey *hsd* test discussed in Chapter 12.

This concludes our discussion of the analysis of variance technique— one of the most useful tools of statistics. Still further extensions allow us to analyze the data from factorial designs involving three or more separate variables. Modifications also have been developed that are appropriate, for example, for analyzing matched groups, such as might be involved in experiments in which a number of measures are obtained from the same subjects rather than from independent groups. Accordingly, the simple and double-classification analysis of variance techniques that we have discussed are the first steps toward the statistical analysis of many experimental problems more complicated than those discussed in this book.

TERMS AND SYMBOLS TO REVIEW

Factorial design

Double classification

Variables

Main effects

Interaction

Additive and nonadditive
    combinations

Sum of squares within groups (SS_{wg})

Sum of squares between groups (SS_{bg})

Sum of squares for group (SS_k)

Sum of squares for interaction $(SS_{A \times B})$

Chi Square

14

Most of the statistical procedures we have discussed in previous chapters have involved *measurement* data. Such data, you recall, involve characteristics that may vary in *amount*. We therefore assign each individual in the group we are measuring a number reflecting *how much* of the characteristic that individual exhibits.

In contrast to measurement data, there are categorical data which involve the assignment of each individual to one of two or more discrete categories and then the determination of the number of cases or frequencies in each category. (This type of data was considered at length in our discussion of probability in Chapter 6.) For example, we might ask each of 200 people chosen at random whether they believe that voting should be made compulsory. In this case we might have only two categories of responses, Yes and No. We would not measure for each individual the "amount" of his response but rather classify his response as belonging to one of the two categories, and for the group as a whole, record the number of people, the frequency, within each category. Note that all of the persons included in each category have the same "score"; there is no variability in the response as classified *within* a category.

OBSERVED AND EXPECTED FREQUENCIES

Typically, the question we want to answer when we have frequency data is *whether the frequencies observed in our sample deviate significantly from some theoretical or expected population frequencies.* We have some hypothesis about what frequencies should be expected—that is, what the population frequencies are. In some cases we may hypothesize equal frequencies in each category, so we want to determine whether the observed frequencies deviate significantly from what would be expected from a hypothesis of equal likelihood. For example, suppose in our voting illustration that 116 of the 200 persons questioned were in favor of compulsory voting and 84 were not. We might want to test the hypothesis that there is a 50:50 split in the population and that the deviation from an equal number of people in each category in our sample was due to sampling error. In other cases we may adopt some other hypothesis. For example, we might hypothesize that the frequencies in the various categories should exhibit some specified ratio. Thus, suppose we knew the number of registered cars in a certain county that were subcompacts, compacts, or standards. These frequencies would exhibit a certain ratio, and we could use them as expected frequencies in testing whether ownership of the three types of cars in a random sample from the whole state deviated significantly from what was true in the county.

In these examples we again have the usual problem of determining whether the deviation of observed values (sample frequencies) from expected values (population frequencies) can be attributed to sampling errors, or whether we can conclude, at a certain level of probability, that a nonchance factor was operating. In our voting example we would not expect that if we measured another random sample of 200 people we again would get exactly 116 Yes and 84 No. We need, therefore, a statistical technique by which we can determine whether or not the frequencies observed in a sample depart significantly from expected frequencies. The statistic we use is called *chi square*, symbolized χ^2.

CHI SQUARE APPLIED TO ONE SAMPLE

The formula for χ^2 is:

$$\chi^2 = \sum \frac{(O - E)^2}{E} \tag{14.1}$$

where: O = observed frequency

  E = the corresponding expected frequency

According to this formula we first get the deviation of each observed frequency from its corresponding expected frequency and then square the deviations. Next, each squared deviation is divided by the appropriate expected frequency (the one used to obtain the deviation). The remaining step is to sum these quotients; the sum is the value of χ^2.

Let us see how this works with our voting example. We shall test the hypothesis that opinion neither favors nor opposes compulsory voting, that in the population from which our sample was drawn opinion is equally divided. The expected frequency is, therefore, 100 in each category. Table 14.1, which shows the computation, indicates that for a 50:50 hypothesis with these data, $\chi^2 = 5.12$. How is this value to be interpreted? The interpretation of a value of χ^2 is basically the same as for t. We assume a certain null hypothesis—for the data in Table 14.1 the hypothesis that there is no difference in the population between the two categories of response. Then we ask: if the hypothesis is true, how often would values of χ^2 as large as or larger than our obtained value arise by random sampling error? If we find that values of χ^2 as large as or larger than the value obtained would occur by sampling error only 5 percent or 1 percent of the time, we reject the hypothesis at the 5% or 1% level of significance.

Table 14.1 A 50:50 hypothesis tested by χ^2.

	Favor compulsory voting	Do not favor compulsory voting	Total
Observed (O)	116	84	200
Expected (E)	100	100	200
O − E	+16	−16	
$(O - E)^2$	256	256	
$\dfrac{(O - E)^2}{E}$	2.56	2.56	

$$\chi^2 = \sum \frac{(O - \ )^2}{E} = 2.56 + 2.56 = 5.12$$

In order to evaluate obtained values of χ^2 we need to know the distribution of χ^2. Table G at the end of the book shows for each *df* value from 1 to 30 the value of χ^2 that would occur by sampling error at both the .05 and .01 levels of probability. Thus, for 1 *df* (first row in Table G) a value of χ^2 as large as 6.64 occurs 1 percent of the time by sampling error alone. Similarly, with 6 *df*, an obtained χ^2 would have to be at least 12.59 to be called significant at the 5% level. Note that, unlike F or *t*, values of χ^2 necessary for significance at a given level *increase* with increasing *df*.

Degrees of Freedom.   The *df* for a particular χ^2 does *not* depend upon the number of individuals in the sample. For χ^2, the *df* is determined by the number of *deviations*, between observed and expected frequencies, that are *independent* (that are free to vary). For example, what is the *df* for Table 14.1? First, we have to note a condition that must always be true for any χ^2: *The sum of the expected frequencies must equal the sum of the observed frequencies.* In Table 14.1 this is true; both observed and expected frequencies sum to 200. Now if, in Table 14.1, we write down 100 as the expected frequency for one category, the expected frequency for the other category (and, therefore, the deviation) is *not* free to vary; it must also be 100, in order that the expected frequencies sum to the same total as the observed frequencies. Therefore, there is 1 *df* for the data in Table 14.1, because only one deviation is free to vary.

We can now use Table G to evaluate the χ^2 for Table 14.1. Table G shows that for 1 *df*, $\chi^2 = 3.84$ at the 5% level, 6.64 at the 1% level. Our obtained value (5.12) exceeds the value at the 5% but not at the 1% level, so we may reject our null hypothesis at the 5% level of significance. We conclude that opinion concerning compulsory voting is quite probably not evenly split in the population from which our sample was drawn.

Chi square may be used to test *any a priori* or *assumed* hypothesis about the population; it is not restricted to testing the hypothesis of equally distributed frequencies. By an a priori or assumed hypothesis we mean one that the investigator has before the research is done; the research is carried out to test the hypothesis. The hypothesis may be based on previous research or be derived from some theory. Suppose, for example, we are trying to get a congressman to introduce a bill making voting compulsory. On the basis of preliminary research or on some other grounds we claim that 75 percent of the voting population favor making voting compulsory. Our hypothesis is a 75:25 split. We shall not carry out the computations for this hypothesis, since the method is already illustrated in Table 14.1. Our expected frequencies would show a 75:25 ratio instead of the 50:50 previously used. Whatever the N in our sample, 75 percent of N would be the expected frequency in the Yes category, 25 percent of N the expected frequency in the No category.

Restrictions on the Use of χ^2

Before taking up the use of χ^2 in more complex cases, we shall note some of the conditions that must be met before data may appropriately be analyzed by χ^2. Like any statistical method, the use of χ^2 is subject to certain limitations; if these are kept in mind, many incorrect applications will be avoided.[1]

First, χ^2 can be used only with frequency data. It is not correct, for example, to test by χ^2 the deviation between the obtained mean score on some trait for a group, and some expected or predicted mean. These are measurement data, and one of the reasons why χ^2 cannot be used with such data is that the value of χ^2 varies with the size of the unit of measurement. Thus, weight may be measured either in pounds or in kilograms. But the value of χ^2 between observed and expected weights will be different, even for the same subjects, if the score is recorded in pounds than if recorded in kilograms. Who is to say which value of χ^2 is "correct"?

Second, the individual events or measures must be independent of each other. The data for the example on compulsory voting may be used as an illustration. There is no reason to think that the answer of any one of the subjects in the sample is dependent upon or correlated with the answer of any other subject; the individual responses are independent. But it would be incorrect, for example, to ask each of fifty people to make guesses as to what card will be drawn from a deck on each of five successive draws and then to claim the total frequency was 250. In such a case we would not have 250

[1] Lewis, D., and Burke, C. J. The use and misuse of the chi-square test. *Psychological Bulletin*, 1949, 46, 433–487.

independent responses; successive responses by the same individual are very probably related.

Third, no theoretical frequency should be smaller than 5. The distribution of χ^2, part of which is given in Table G, may take different values from those shown in the table if any expected frequency is less than 5. The discrepancy is not large, however, when χ^2 is computed from contingency tables with a fairly large number of cells (more than 4, at a minimum) and only a few theoretical frequencies are less than 5. In this latter instance, it is permissible to relax the stringent rule that *no* theoretical frequency may be less than 5 and to go ahead and compute χ^2.

Fourth, there must be some basis for the way the data are categorized. Preferably, the categories should have some logical basis, or be based on previous acceptable research, and should be set up before the data are collected. Suppose, for example, we asked a large random sample of people what magazine they would take if they could subscribe to only one. We find that thirty-one magazines were named, ranging from a few mentioned very frequently down to several preferred by one person each. We now categorize the data into several "types" of magazines in order to apply χ^2. Such an application would be questionable, because we have not demonstrated that the types (categories) we set up are any more defensible than any others that might be used. And it would be even worse to continue reclassifying the data until we got categories that would give a significant χ^2. Our categories must be set up beforehand, and it must be demonstrated that the categories are reliable. Reliability would be demonstrated if there were a high degree of agreement among a group of qualified judges as to what magazines should be assigned to what categories.

Finally, recall that we said above that *the sum of expected and the sum of observed frequencies must be the same.* This requirement is related to another: *if we are recording whether or not an event occurs, we must include in our data, and use in computing χ^2, both the frequency of occurrence and the frequency of nonoccurrence.* For example, suppose we roll a die 120 times and note the frequency with which a one-spot turns up. The expected frequency is 20, since a die has six sides. It would not be correct to test by χ^2 only whether the observed frequency of occurrence of the one-spot deviated significantly from the expected frequency. The χ^2 must be based *both* on the deviation of observed frequency of occurrence from expected frequency of occurrence, and the deviation of observed frequency of nonoccurrence from expected frequency of nonoccurrence. If we ignore the frequency of non-occurrence, the *sums* of observed and expected frequencies will not be the same, except by chance. The value of χ^2 must be based on the deviation of observed from expected frequency for both the category of occurrence and the category of nonoccurrence.

TESTING SIGNIFICANCE OF
DIFFERENCE BY χ^2

Chi square probably has its greatest usefulness in testing for significance of differences between groups. Although there may be any number of groups and any number of categories, most often in research we have two groups and two categories of response; the data are expressed in a 2×2 table. Our illustration of χ^2 as it is used to test for significance of difference therefore will be based on a 2×2 table, although the method is the same for any number of groups and categories.

Suppose we want to determine whether there is any relation between the amount of education a person has had and his method of keeping up with the news. We ask a random sample of college graduates and a random sample of high school graduates whether they keep up with the news mostly by reading a newspaper or mostly by watching television. Table 14.2 shows the data that might have been obtained, arranged in a 2×2 table.

Table 14.2 shows that 47 of the 109 college graduates in the sample said they got news mostly by watching television; the other 62 relied mostly on the newspaper. Among the 97 high school graduates, 58 said television, 39 the newspaper. We want to know whether these frequencies indicate a significant difference between the two groups.

When χ^2 is used to test for the significance of a difference between two or more groups, it is sometimes said to be a test of *independence*. Perhaps this can best be explained by pointing out that the observed frequencies in Table 14.2 are classified two ways. One way is television versus newspaper; the other way is by educational level. Our problem, stated above in terms of significance of difference, can therefore be restated as the question: Are the

Table 14.2 Testing significance of difference by χ^2.

	Television	Newspaper	Total
College graduates	47(55.6)	62(53.4)	109
High school graduates	58(49.4)	39(47.6)	97
Total	105	101	206

O	E	O − E	$(O - E)^2$	$\dfrac{(O - E)^2}{E}$
47	55.6	−8.6	73.96	1.33
62	53.4	+8.6	73.96	1.39
58	49.4	+8.6	73.96	1.50
39	47.6	−8.6	73.96	1.55
				$\chi^2 = 5.77$

two ways of classifying the observed frequencies independent of each other? If they *are* independent, then preference for television or newspaper does *not* depend on educational level and we could combine the educational groups, categorizing merely on the basis of television versus newspaper. This is the same as saying the groups do *not* differ significantly. On the other hand, if the classifications are *not* independent—that is, if they are correlated—then when we categorize the data one way, we must categorize the other way too. In our example this would mean that the educational groups *differ significantly* in their preference for television or newspaper.

Let us now see how we use χ^2 to test for significance of difference, using our example in Table 14.2. We start by assuming the null hypothesis that our two groups were drawn from the same population with respect to method of keeping up with the news. Note that we are not concerned, for either group, with whether television is preferred over the newspaper. If we were, we could use χ^2, as explained earlier, to test a 50 : 50 hypothesis in each group separately.

Our next step is to determine the expected frequency for each of the four cells of the table. Since we have no a priori hypothesis, we reason as follows: if our null hypothesis is correct—if the true (expected) frequencies are the same for both samples—then combining the two samples should give us a better estimate of the true frequencies than we could get from either sample alone. Thus we add, within each column, the observed frequencies for the samples to get the marginal column totals of 105 and 101. We use these values as the best estimates of the division in the single assumed population between television preference and newspaper preference. Since our total sample is 206 cases, the expected percent frequency in the population for television preference is $^{105}/_{206}$, or 50.97 percent. In other words, our best estimate of what we would obtain if we measured the whole population is 50.97 percent getting news mostly by television. In similar fashion, $^{101}/_{206}$, or 49.03 percent of the total frequency, is the expected percent frequency for the newspaper category. All we have to do now, to get the expected raw frequencies for each sample, is divide the sample N on the basis of these expected percent frequencies. Thus, there are 109 college graduates; since we expect 50.97 percent of them to fall in the television category, we have 50.97 percent of 109, or 55.6, as the expected frequency for the upper left-hand cell of the table. This value is shown in parentheses in that cell in Table 14.2. Similarly, 50.97 percent of the 97 high school graduates gives us 49.4 as the expected frequency in the television category for those subjects. The expected frequencies for the newspaper cells are determined by taking 49.03 percent of 109 and of 97.

Now that we have stated the logic for the method of determining the expected frequencies, we can state a simple rule that may be more easily remembered: simply multiply the marginal total for any column by the marginal total for any row and divide the product by the total N. The result is the expected frequency for the cell common to the row and column whose totals were multiplied together. Thus, in our example, $(101)(97)/206 = 47.6$; $(105)(109)/206 = 55.6$, and so forth.

Returning to our problem, now that we have the expected frequencies, we compute χ^2 in the usual manner: each deviation of an observed frequency from its expected frequency is squared, divided by the expected frequency, and the quotients summed. We find that $\chi^2 = 5.77$ for this example. Now, what are the *df*? We shall state a general rule for determining the *df* for any table that has at least two rows and two columns, and in which the marginal totals are used in determining the expected frequencies. The *df* equals (number of columns $-$ 1)(number of rows $-$ 1). Thus in a 2 $\times$ 2 table, the $df = (2 - 1)(2 - 1)$, or 1. The rule applies to tables of any number of rows and columns, as long as there are at least two of each.

With 1 *df*, our χ^2 value of 5.77 is larger than the value given in Table G for the 5% level. We shall therefore reject the hypothesis at the 5% level of significance that the two groups were drawn from the same population with respect to method of getting the news. We could say the same thing by stating that we reject the hypothesis of independence; the frequencies in the television versus newspaper categories are probably *not* independent of educational level.

Before we leave this example we may note that χ^2 can be computed for a 2 $\times$ 2 table without computation of expected frequencies.[2] Let us schematize a 2 $\times$ 2 table as follows, where the letters in the cells indicate the *observed* frequencies:

A	B	A + B
C	D	C + D

A + C   B + D   N

With this schema, χ^2 can be computed directly as follows:

$$\chi^2 = \frac{N(AD - BC)^2}{(A + B)(C + D)(A + C)(B + D)} \qquad (14.2)$$

[2] Reprinted with permission from McNemar, Q., *Psychological Statistics*, New York, John Wiley & Sons, Inc., 1949.

Thus for the data in Table 14.2:

$$\chi^2 = \frac{206[(47)(39) - (62)(58)]^2}{(109)(97)(105)(101)}$$

Significance of Differences Among Several Groups

The method of applying χ^2 to test for significance of differences among several groups is identical with the method illustrated in the previous section with two groups. As an example, consider the data in Table 14.3. The data represent the number of bachelor's degrees awarded by a university in 1955, 1965, and 1975 in four different fields: social sciences, natural sciences, humanities, and education.

Let us apply χ^2 to the data in Table 14.3. We shall be testing for any overall significance of differences among the groups. The procedure is the same as that previously illustrated with a 2×2 table. We obtain the

Table 14.3 Testing overall significance of difference among several groups by χ^2.

Degree	1955	1965	1975	Total
Social Sciences	46	67	195	308
Humanities	51	53	88	192
Natural Sciences	85	100	154	339
Education	118	163	280	561
Total	300	383	717	1400

O	E	O − E	(O − E)²	$\frac{(O - E)^2}{E}$
46	66.00	−20.00	400.00	6.06
51	41.14	9.86	97.22	2.36
85	72.64	12.36	152.77	2.10
118	120.21	−2.21	4.88	.04
67	84.26	−17.26	297.91	3.54
53	52.53	.47	.22	.00
100	92.74	7.26	52.71	.57
163	153.47	9.53	90.82	.59
195	157.74	37.26	1388.31	8.80
88	98.33	−10.33	106.71	1.09
154	173.62	−19.62	384.94	2.22
280	287.31	−7.31	53.44	.19
			$\chi^2 =$	27.56

theoretical frequency for each cell by multiplying the total for the row containing the cell by the total for the column containing the cell and dividing by the total N, which here is 1400. For example, for the cell in the row for Social Sciences and the column for the year 1955 we have 308(300)/1400 = 66.00. In Table 14.3 the theoretical frequencies are shown in the E column. Then χ^2 is computed in the usual way: in each cell subtract the theoretical frequency from the observed frequency to get the deviation (O − E), square the deviation, and divide by the theoretical frequency of the cell; the sum of the quotients is the value of χ^2. For the data in Table 14.3, $\chi^2 = 27.56$. Since this is a 3 × 4 table, there are 6 *df*: (3 − 1)(4 − 1). Looking at Table G, we see that for 6 *df* a χ^2 of 16.81 is required for significance at the 1% level. Our obtained χ^2 is far greater than this value. We therefore reject the hypothesis that the proportion of degrees awarded was constant in the three years for which data were collected. (Inspection of Table 14.3 suggests that social sciences and education were gaining in popularity while the natural sciences and, even more, the humanities were losing.)

Chi Square with More than 30 *df.*   Occasionally we may have so many groups and categories that there are more than 30 *df*. This would be the case in a 6 × 8 table (35 *df*), a 9 × 9 table (64 *df*), and so on. Table G does not go beyond 30 *df*. To evaluate a χ^2 based on more than 30 *df*, we convert the χ^2 value to an x/σ or z score value and look up the probability in the normal curve table (Table B). The following expression is used to convert a χ^2 based on more than 30 *df* to an x/σ value:

$$\frac{x}{\sigma} = \sqrt{2\chi^2} - \sqrt{2n - 1} \qquad\qquad (14.3)$$

where n = the number of *df* for the χ^2 value.

Correction for Continuity

The values of χ^2 presented in Table G were obtained from theoretical distributions of χ^2, the exact shape of each distribution (and hence the exact values associated with the 5% and 1% levels) being determined by *df*. Each of these theoretical distributions involves continuous rather than discrete values of χ^2. That is, χ^2 does not jump in value from, for example, 2.0 to 3.0 or from 16.5 to 17.5, but may take *any* fractional value between any pair of adjacent integers. Calculated values of χ^2, on the other hand, do vary in discrete fashion, since they are computed from frequencies that are always whole numbers and thus never take fractional values. This fact creates a problem in interpreting χ^2, particularly when *df* and expected frequencies are small, but the problem can be at least partially overcome by introducing what is called a *correction for continuity* into our computational steps when determining the χ^2 for a set of data. The correction procedure is a simple one. After we determine the O − E values, we merely reduce the *absolute* magnitude of each of them by .5. For example, an O − E of −1.65 would be reduced to an absolute value of 1.15 and an O − E of .87 to .37. We then square each of these *corrected* O − E values and proceed from there as usual.

Introduction of the correction factor into the computational procedures results in relatively little change in χ^2 unless both *df* and the expected frequency in one or more cells are small (which reflects the fact that discontinuities in the distribution of obtained χ^2's are marked only in these cases of small *df* and expected frequencies). A general rule therefore adopted by statisticians is that it is necessary to apply the correction for continuity only when *df* = 1 (as in a 2 × 2 or 1 × 2 table) *and* the expected frequency in one or more of the cells is less than 10. Note carefully that it is not the value of the *observed* frequency in a cell but rather the value of the *expected* frequency that must be less than 10 for us to require that the correction for continuity be used.

We will now illustrate the application of the correction for continuity, using the data in Table 14.4. The data represent the results of an experiment investigating the effectiveness of a new drug in preventing a certain virus infection. An experimental group of 25 animals was given the drug and a control group of 20 animals no drug; all animals were then inoculated with the virus and later examined for presence of the infection. Assuming that we have no a priori hypothesis about what proportion in either group will develop the infection after inoculation and wish only to determine whether

Table 14.4 Chi-square test of independence with correction for continuity applied to data of drug study.

	Infection	*No infection*	*Total*
Experimental	11(14.44)	14(10.56)	25
Control	15(11.56)	5(8.44)	20
Total	26	19	45

O	E	O − E	$(O - E)_c =$ $(\lvert(O - E)\rvert - .5)$	$(O - E)_c^2$	$\dfrac{(O - E)_c^2}{E}$
11	14.44	−3.44	2.94	8.64	.60
15	11.56	3.44	2.94	8.64	.75
14	10.56	3.44	2.94	8.64	.82
5	8.44	−3.44	2.94	8.64	1.02
				$\chi^2 =$	3.19

the drug group is more resistant to infection than the controls, we proceed to apply the χ^2 test of independence. In the 2 × 2 table of Table 14.4 we have presented the number of animals in each group that did and did not develop the infection (observed frequencies) and in parentheses the expected frequencies that we have calculated for each of the four categories. (The expected frequency for each cell, you will recall, is found by multiplying the marginal row and column totals for that cell and dividing the product by the total N.) We can easily observe that the data are such that the correction for continuity should be used in computing χ^2, since $df = 1$ and one of the expected frequencies is less than 10. At the bottom of the table we have corrected each O − E by taking its *absolute* value, as indicated by the expression ($|O - E|$), and reducing it by .5. We then have determined χ^2 using these corrected figures. The χ^2 of 3.19 we obtained falls short of 3.84, the value of χ^2 we find in Table G to be required at the 5% level with 1 *df*. We therefore accept the null hypothesis and conclude that the control and experimental groups did not differ significantly in the proportion of animals who developed the infection. If we had computed χ^2 for these data without the correction factor, we would have found χ^2 to be 4.36, a value that would lead us to *reject* the null hypothesis at the 5% level. Failing to correct for continuity, then, might well have led us to reject a true hypothesis.

The formula we presented earlier for calculating χ^2 for a 2 × 2 table without computation of the expected frequencies can also be modified to include a correction for continuity. This modified formula is:

$$\chi^2 = \frac{N(|AD - BC| - N/2)^2}{(A + B)(C + D)(A + C)(B + D)} \tag{14.4}$$

where the vertical lines on either side of the expression AD − BC, as well as in Table 14.4 above, indicate "without regard to sign." Thus, the effect of subtracting the correction factor, N/2, is to reduce the absolute size of the AD − BC term. Use of the formula with the data from our drug experiment illustrates that (except for rounding error) it produces the same answer as we calculated in Table 14.4 Thus:

$$\chi^2 = \frac{45(|(11)(5) - (14)(15)| - 45/2)^2}{(25)(20)(26)(19)}$$

$$= \frac{45(|-155| - 22.5)^2}{247,000}$$

$$= \frac{45(132.5)^2}{247,000}$$

$$= 3.20$$

Let us conclude by reemphasizing the importance of the restrictions on the use of χ^2 that we discussed. These restrictions are not mere hairsplitting, since the use of χ^2 when it is inappropriate may lead us to erroneous conclusions about our data. The same statement can be made about any statistical test; we must be sure before applying a statistical test to any set of data that the assumptions underlying its use are not violated.

In the next chapter we will consider several additional statistical techniques that may be appropriate when the assumptions of some of the tests we have discussed so far cannot be met.

TERMS AND SYMBOLS TO REVIEW

Frequency data

A priori hypothesis

Observed frequency (O)

Test of independence

Expected frequency (E)

Correction for continuity

Chi square (χ^2)

Nonparametric Tests

Tests

15

When we have measured two or more samples, we may determine the significance of the difference between or among their $\overline{X}$'s by means of a t test or of the F in an analysis of variance. These statistical techniques are examples of what are called *parametric* tests, since they require that we estimate from the sample data the value of at least one population characteristic (parameter) such as its SD. One of the assumptions we make in applying these parametric techniques to sample data is that the variable we have measured is normally distributed in the populations from which the samples were obtained.

In this chapter we will consider several of the *nonparametric* techniques statisticians have devised, techniques whose theory makes no use of parameter values. A second attribute of these nonparametric methods is that they are based on less restricting assumptions than those underlying parametric ones concerning the shape of the distribution of the characteristic being measured. Thus an alternate label for these methods is *distribution-free*.

NONPARAMETRIC VERSUS PARAMETRIC TESTS

Quite obviously, nonparametric techniques become very valuable when the samples we have measured suggest that the assumptions underlying an otherwise appropriate parametric test cannot be met. Nonparametric tests, however, can be applied not only to measurement data but also to other types such as sets of ranks or classified frequencies. For example, we might ask a factory foreman to rank twenty men working under his supervision, from the best to the poorest worker, and then compare the ranks received by ten of the men who had been given a training course before being assigned to their job with the ranks of the ten men who had been given no special training. It is unnecessary to give an illustration of classified frequencies, since we have just dealt with such data in Chapter 14 on χ^2. We should mention, however, that χ^2 is itself an example of a nonparametric technique and that several of the tests we will be discussing in this chapter involve converting our raw data into classified frequencies and then computing a χ^2.

When we have measurement data for which it is permissible to apply a parametric test, we often find it more time consuming and computationally involved to compute the t, analysis of variance, or whatever, than a parallel nonparametric test. However, whenever possible we continue to use parametric tests in preference to nonparametric ones for the following reason.

If we apply both to the same data, we usually find that a nonparametric test is less powerful in terms of rejecting the null hypothesis: other things being equal, a larger sample difference is needed between groups to reject the null hypothesis at a given significance level for a nonparametric than for a parametric test. Since, as research workers, we are particularly concerned with minimizing the probability that we will accept our null hypothesis when it is false, we use the more powerful parametric tests whenever we can. Nonparametric techniques, then, are reserved for situations in which we need the results of a significance test in a hurry or in which the nature of our data makes the use of a parametric test inappropriate. The latter occurs either because we are not dealing with measurement data or because our measurement data suggest that the assumptions underlying the use of parametric tests are not met and that important errors may result from their use.

SOME NONPARAMETRIC TESTS *a broad measure*

Median Test with Two Independent Samples

The first nonparametric procedure we will consider, the *median test,* is appropriate when we wish to compare two independent, random samples. With this technique we test the null hypothesis that the two samples were drawn from populations with the same *median.*

In conducting the test, we first combine the two samples and find the median of the total, combined group. Then we count the number of cases in each sample that fall above and below the overall median. A problem that may arise in this process is caused by the occurrence of scores that take the same value as the median. If an even number of scores in a given sample takes this value, half should be placed in the above-the-median category and the other half in the below category. If an odd number of scores occurs, the "extra" score will have to be placed in one of the two categories, chosen at random.

To illustrate these initial steps, let us assume we have two independent samples with the following scores:

Sample 1 (N = 20)

6	10	3	4	8	2	9	2	7	4
5	3	6	1	9	11	4	8	9	3

Sample 2 (N = 22)

3	15	11	9	14	15	11	4	10	12	15
14	11	12	6	13	6	12	7	14	8	15

When all 42 cases are pooled, the median for the total group is found to be 8.5. When we count the number of cases in each sample that are above and below the median of 8.5, we find:

	Above median	*Below median*	*Total*
Sample 1	5	15	20
Sample 2	16	6	22
Total	21	21	42

Now if the samples came from populations with the same median, we would expect that except for sampling error, half of the cases in each sample would fall above the median of the total group and half below it. To determine whether there is a significant departure from a 50:50 split within each of our samples, we now perform a χ^2 test with $df = 1$, following the procedures outlined in the last chapter (including, whenever appropriate, the application of the correction for continuity.)

In Table 15.1 we have computed the χ^2 value for the data presented above. Since we were testing a 50:50 hypothesis, we obtained the expected frequencies for the cases above and below the median in each sample simply by dividing the sample N by 2. The χ^2 we found, 9.54, is greater than the value of 6.64 required for significance at the 1% level with 1 df. We therefore reject the null hypothesis and conclude that the samples came from populations with different medians.

Table 15.1 Median test: χ^2 testing the 50:50 hypothesis for two groups.

	Above median	*Below median*	*Total*
Sample 1	5(10)	15(10)	20
Sample 2	16(11)	6(11)	22
	21	21	

O	E	O − E	(O − E)²	$\frac{(O - E)^2}{E}$
5	10	−5	25	2.50
16	11	5	25	2.27
15	10	5	25	2.50
6	11	−5	25	2.27
				$\chi^2 = 9.54$

Median Test with More than Two Independent Samples

The median test we have just described may actually be used to compare any number of independent samples, not merely two. To review our method, the k samples are first combined and the median for the total group of cases determined. We then set up a $2 \times k$ table, entering the number of cases in each of the k samples that fall above and below the overall median. Using these data, we then compute a χ^2 testing the hypothesis that there is a 50:50 split in each of the samples. In interpreting this χ^2 we use $df = (2 - 1)(k - 1)$, which, of course, reduces to $k - 1$, or one less than the number of samples. If our computed χ^2 exceeds the tabled χ^2 value for $df = k - 1$ at the 5% or 1% level, we reject the null hypothesis that the samples all came from populations with the same median.

A Rank Test for Two Independent Samples (Mann-Whitney U)

The U test, devised by Mann and Whitney and also, independently, by Wilcoxon, is a technique based on score ranks that is appropriate for comparing two independent groups. Our first step is to combine the two samples and rank the total set of scores, assigning the rank of 1 to the largest score, 2 to the second largest, and so on. (The same final result would be obtained if scores were instead ranked from *smallest* to largest.) If tied scores occur, we handle them in the usual way: each is assigned the average of the ranks the tied scores jointly occupy; for example, if two scores are tied for fourth place, each is assigned a rank of 4.5, the average of 4 and 5.[1]

Having assigned a rank to each score, we find the sum of the ranks for each sample, identifying these sums by R_1 and R_2. Next we calculate a statistic designated as U, one for each sample. (Actually, we need only one of these U values but can use the other as a check on the accuracy of our calculations.) For Sample 1, this statistic is determined by the following formula:

$$U_1 = N_1 N_2 + \frac{N_1(N_1 + 1)}{2} - R_1 \qquad (15.1)$$

where R_1 is the sum of the ranks assigned to the first sample.

[1] No problem results from this method of assigning ranks to tied scores when they all occur in a single sample, since the sum of ranks is unaffected. When ties occur across groups, the sum for each group *is* affected, being determined by the particular rank given to each tied score. This, in turn, affects U_1 (or U_2) and z. One statistician, Bradley, J. V., *Distribution-Free Statistical Tests* (Englewood Cliffs, N.J.: Prentice-Hall, Inc.), 1968, has recommended that this problem be resolved by considering all possible rankings of tied values and computing and reporting the z's based on the U's of all the various combinations. This recommendation seems reasonable in instances in which the conclusion one reaches about the significance of the findings is likely to be affected.

Assuming the null hypothesis, which states, in effect, that the samples come from the same population with respect to the measured characteristic, the sampling distribution of U has a mean (expected U or U_E) equal to the value given by the formula:

$$U_E = \frac{N_1 N_2}{2} \tag{15.2}$$

and a standard deviation whose value is given by the formula:

$$\sigma_U = \sqrt{\frac{N_1 N_2 (N_1 + N_2 + 1)}{12}} \tag{15.3}$$

We now convert our obtained U into a z score:

$$z = \frac{x}{\sigma} = \frac{U_1 - U_E}{\sigma_U} \tag{15.4}$$

The U for the second sample could be obtained by a parallel formula, with R_1 being replaced by R_2 and, in the second term of the formula, the N_1's by N_2's. We could then find z by subtracting U_E from U_2 instead of U_1 in the numerator of the formula, since the two resulting differences are opposite in sign but of the same absolute value; that is, U_1 and U_2 deviate equally from U_E but in the opposite direction. The scores of the sample whose U is less than U_E are, of course, lower than those of the other sample and vice versa.

If the N of each of our samples is 8 or more, the sampling distribution of U is normal or approximately so, and we may use the normal curve table to assess the null hypothesis. Thus, if we are testing a bi-directional H_1 and the z we have calculated is greater than 1.96 (or 2.58), we reject the null hypothesis at the 5% (or 1%) level and conclude that the samples differed significantly. If z is less than this value, we accept the null hypothesis. When the N of one or both of the samples is less than 8, the sampling distribution of U is not normal and we must use a special table, which is not included in this book, to interpret z.

In Table 15.2 we have performed a Mann-Whitney U test on the hypothetical data from two independent samples. The z has been calculated twice, first using U_1 and then U_2, to demonstrate that z will be the same except for sign. Since the sample N's are greater than 8, we use the normal curve to assess our calculated z of 1.34 and see that it falls far short of 1.96, the value needed for significance at the 5% level. We therefore accept the null hypothesis and conclude that the two samples could have come from the same population.

Table 15.2 Illustration of the calculation of the Mann-Whitney U test*.

Sample 1 (N = 11)		Sample 2 (N = 10)	
X	Rank	X	Rank
35	1	34	2
33	3	31	5
32	4	26	9
30	6	25	11
27	7	23	13
26	9	22	14
26	9	21	15
24	12	18	19
20	16.5	17	20
20	16.5	16	21
19	18		$R_2 = \overline{129}$
	$R_1 = \overline{102.0}$		

$$U_1 = N_1 N_2 + \frac{N_1(N_1 + 1)}{2} - R_1$$

$$= (11)(10) + \frac{11(11 + 1)}{2} - 102$$

$$= 176 - 102 = 74$$

or

$$U_2 = N_1 N_2 + \frac{N_2(N_2 + 1)}{2} - R_2$$

$$= (11)(10) + \frac{10(10 + 1)}{2} - 129$$

$$= 36$$

$$U_E = \frac{N_1 N_2}{2} = \frac{(11)(10)}{2} = 55$$

$$\sigma_U = \sqrt{\frac{N_1 N_2 (N_1 + N_2 + 1)}{12}}$$

$$= \sqrt{\frac{(11)(10)(11 + 10 + 1)}{12}}$$

$$= \sqrt{201.67} = 14.20$$

$$z = \frac{U_1 - U_E}{\sigma_U} \quad \textbf{or} \quad z = \frac{U_2 - U_E}{\sigma_U}$$

$$= \frac{74 - 55}{14.20} = \frac{19}{14.20} \qquad = \frac{36 - 55}{14.20} = \frac{-19}{14.20}$$

$$= 1.34 \qquad = -1.34$$

* Each listed rank is the overall rank of the associated X in the total group of twenty-one cases.

In both our discussion and illustration of the Mann-Whitney U test we have assumed that the raw scores were measurement data of some kind and that we first had to convert them into ranks in order to perform the test. Sometimes, however, the raw scores *are* the ranks we need, as in the example given earlier in which the work efficiency of twenty men had been ranked, and we wished to compare the efficiency of those in the total group who had been given a special training course with the efficiency of those who had not. Except for omitting the initial step of converting raw scores to ranks, we would compute the Mann-Whitney U in the usual manner with data of this type. The same remarks hold true for the other rank tests we will discuss.

A Rank Test for Two or More
Independent Groups (Kruskal-Wallis One-Way
Analysis of Variance by Ranks)

The Kruskal-Wallis one-way analysis of variance by ranks is an extension of the Mann-Whitney U test that allows us to compare any number of independent groups rather than only two. We first pool all of the k groups, arrange the total group of scores in order, and assign the *smallest* score a rank of 1, the next smallest a rank of 2, and so on, handling ties in the usual manner. (It is essential in this test that the *smallest* score be given a rank of 1, and so on down to the *largest* score.) Having ranked the total group, we then find the sum of the ranks for each of the k samples (R_i).

If each of the samples came from the same population, then each of these sums of ranks (each R_i) should be equal except for sampling error. We test this null hypothesis by computing a statistic H by the following formula:

$$H = \frac{12}{N(N+1)} \sum \left(\frac{R_i^2}{n_i} \right) - 3(N+1) \tag{15.5}$$

where: N = total number of cases in the k groups

n_i = number of cases in a given sample

R_i = sum of the ranks in a given sample

The expression $\sum (R_i^2/n_i)$, then, indicates that, for each group, we are to square the sum of ranks (R), divide this R^2 by the number of cases in the group, and finally to sum these quantities for the k groups. An illustration of these procedures involving four groups is shown in Table 15.3.

When each of the samples has five or more cases, the statistic H takes the chi-square distribution with df equal to $k-1$, the number of groups minus one. In our example where $k=4$, therefore, we enter Table G with $4-1$ or 3 df and find that a χ^2 of 11.34 is required for significance at the 1% level. The H of 22.93 we obtained in Table 15.3 is greater than this value, so we are able to reject the hypothesis that all of our samples came from the same population.

Table 15.3   Illustration of the calculation of the Kruskal-Wallis one-way analysis of variance for ranks*.

Group I (n = 6)		Group II (n = 8)		Group III (n = 7)		Group IV (n = 7)	
X	Rank	X	Rank	X	Rank	X	Rank
99	28	85	18.5	75	9	91	23
98	27	82	16	70	6	89	22
96	26	81	14	68	5	86	20
95	25	81	14	67	4	85	18.5
93	24	78	11	65	3	84	17
88	21	76	10	62	2	81	14
	$R_1 = 151$	73	8	59	1	79	12
		72	7		$R_3 = 30$		$R_4 = 126.5$
			$R_2 = 98.5$				

$$H = \frac{12}{N(N + 1)} \Sigma \left(\frac{R_i^2}{n_i} \right) - 3(N + 1)$$

$$= \left[\frac{12}{28(28 + 1)} \right] \left[\frac{151^2}{6} + \frac{98.5^2}{8} + \frac{30^2}{7} + \frac{126.5^2}{7} \right] - 3(28 + 1)$$

$$= (.0148)(7427.56) - 87 = 22.93$$

* Each listed rank is the overall rank of the associated X in the total group of twenty-eight cases.

A Rank Test for Two Matched Samples (Wilcoxon Signed-Ranks)

The nonparametric techniques we have discussed so far are used to compare two or more *independently* selected groups. The test we take up now was designed to allow us to compare distributions consisting of matched pairs: the same individual tested under two conditions, or pairs of different individuals matched on some basis of similarity before being tested. The procedure, devised by Wilcoxon, is identified as a matched-pairs signed-ranks test or, more simply, as the Wilcoxon signed-ranks test.

To illustrate, assume we have tested each member of a group under two different experimental conditions. Our first step is to find the difference (D) between the pair of scores (I and II) for each subject. We now discard from further consideration any pair of scores for which the D is 0. We then rank the remaining D's in order of *absolute* magnitude, assigning the rank of 1 to the *smallest* absolute D, 2 to the next *smallest*, and so on. As always, ties are handled by assigning to each tied D the average of the ranks the ties jointly occupy. Next we give to each of these absolute ranks the sign of the associated D score, giving us a set of *signed-ranks*.

These procedures are illustrated in Table 15.4, in which, for convenience, we ordered the ten subjects in terms of the absolute magnitude of their D scores, shown in the fourth column. Since the first subject received the same score in both conditions, so that D = 0, no rank was assigned to the D for this individual. For the nine remaining D's the *smallest* absolute D was given a rank of 1, and so on, down to rank 9 for the largest absolute D. Finally, in the last column we have given each of the absolute ranks the same sign, + or −, as its D.

Our next step is to sum the positive signed-ranks and then to sum the negative signed-ranks. If the two distributions of scores come from the same population—that is, if the null hypothesis that our two experimental conditions have no differential effects is correct—we would expect that the positive and negative sums would have the same absolute value except for sampling error.

For small samples a special table has been set up that allows us to test the null hypothesis directly from the sums of our signed-ranks. This table is reproduced in Table H at the end of the book for N's up to 25, where N is the *total number of signed-ranks* (N_{S-R}). Since some of our D's may be 0 and thus discarded, N_{S-R} may be less than the total number of original pairs of scores. The table is used as follows. The *smaller* (in absolute

Table 15.4 Illustration of the calculation of the Wilcoxon signed-ranks test*.

Subject	Score I	Score II	D	Rank of $|D|$	Signed-Rank
1	39	39	0	—	—
2	47	48	−1	1	−1
3	29	27	2	2.5	2.5
4	33	35	−2	2.5	−2.5
5	36	33	3	4	4
6	48	43	5	5	5
7	38	31	7	6.5	6.5
8	43	36	7	6.5	6.5
9	32	23	9	8	8
10	52	40	12	9	9
					$\sum+ = \overline{41.5}$
$N_{S-R} = 9$		$T = 3.5$			$\sum- = -3.5$

* The raw data are from 10 subjects tested under two experimental conditions.

value) of the two sums of signed-ranks is designated as T. We then look up in Table H the *maximum* value T may take and still be significant at the 5% or 1% level (two-tailed test) for our N_{S-R}. (Unlike t or F, the smaller the T the more significant it is.) In our example in Table 15.4 the number of signed-ranks is 9 and T is 3.5. Looking at Table H, we find that for $N_{S-R} = 9$, a T of 6 *or less* is required for significance at the 5% level and a T of 2 *or less* at the 1% level. Since our obtained T of 3.5 is less than 6 but not less than 2, we reject the null hypothesis at the 5% level.

For larger samples the sampling distribution of T is approximately normal with a mean (expected T or T_E) given by the formula:

$$T_E = \frac{N(N + 1)}{4} \tag{15.6}$$

and a standard deviation given by the formula:

$$\sigma_T = \sqrt{\frac{N(N + 1)(2N + 1)}{24}} \tag{15.7}$$

where N = the number of *signed-ranks*. We can then determine the z for our calculated T:

$$z = \frac{T - T_E}{\sigma_T} \tag{15.8}$$

We assess the null hypothesis by comparing our calculated z with the values of 1.96 and 2.58 required for significance in a normal distribution at the 5% and 1% levels, respectively.

A Rank Test for Two or More Matched Groups (Friedman Two-Way Analysis of Variance by Ranks)

On occasion we might want to test the same subjects not merely twice but three or more times. We might, for example, give each of a group of students three passages to memorize that contain political opinions with which they agree strongly, disagree strongly, or have no reactions, in order to determine whether this content factor affects their memorization. Or we might train a group of animals to run to a 10-unit food reward, then give twenty trials with a 5-unit reward to see if there is any systematic decline in running speeds to the reduced reward over the twenty trials. If the underlying assumptions are met, the data obtained by repeated measurement of the same subject may be analyzed by extensions of the analysis-of-variance techniques discussed in earlier chapters—extensions that will not be presented in this text.

If the assumptions for the analysis of variance are not fulfilled by our data, a nonparametric test that may be applied to data involving two or more measures from each subject is the Friedman two-way analysis of variance for ranks. In this analysis one variable is considered to be subjects and the other variable is measures.

In applying the Friedman rank test we first set up a table such as the one at the upper left of Table 15.5, in which each of the rows contains the k scores earned by each of the N subjects. In our example each subject was measured three times. As shown at the upper right of Table 15.5 we now rank, from 1 to k, the scores in each *row*—the scores of each *subject*. We have chosen in our example to go from largest to smallest in ranking the scores of each subject but could have done it the opposite way if we wished.

The next step is to find the sums of the ranks for each of the k measures—that is, for each of the *columns*. If our experimental conditions had no differential effects, any difference in the rank of the scores for a given

Table 15.5 Illustration of the calculation of χ_R^2 in the Friedman two-way analysis of variance by ranks*.

| | Raw scores | | | | Scores ranked by rows | | |
| | Measure | | | | Measure | | |
Subject	I	II	III	Subject	I	II	III
1	12	14	11	1	2	1	3
2	13	19	16	2	3	1	2
3	15	18	19	3	3	2	1
4	17	22	20	4	3	1	2
5	17	15	22	5	2	3	1
6	19	21	24	6	3	2	1
7	21	19	18	7	1	2	3
					17	12	13

$$\chi_R^2 = \frac{12}{Nk(k+1)} \left(\sum R_i^2\right) - 3N(k+1)$$

$$= \left[\frac{12}{(7)(3)(3+1)}\right] [17^2 + 12^2 + 13^2] - 3(7)(3+1)$$

$$= 86 - 84 = 2 \qquad df = k - 1 = 2$$

* The raw data are from seven subjects tested under three experimental conditions.

subject is due to chance factors, and across subjects, any given rank is therefore as likely to be found with one measure as another. This implies, of course, that if the null hypothesis is true, each of the column sums of ranks would be equal except for sampling error.

We test the null hypothesis by computing a χ^2 based on the column sums. This χ^2 is given by the formula:

$$\chi_R^2 = \frac{12}{Nk(k+1)} (\sum R_i^2) - 3N(k+1) \tag{15.9}$$

where: k = number of measures

   N = number of subjects

   R_i = a column sum of ranks

The resulting χ_R^2 approximates the chi-square distribution with $df = k - 1$. In Table 15.5 we calculated χ_R^2 to be 2.00 and df to be 2. This value is far less than the χ^2 of 5.99 found in Table G to be associated with the 5% level for 2 df. We therefore accept the null hypothesis and conclude that there was no significant difference among the three measures.

RELATIVE EFFICIENCY OF PARAMETRIC AND NONPARAMETRIC TESTS

We stress once more that a nonparametric test may not be as powerful as the corresponding parametric test such as t or F. One measure of the relative usefulness of these types of tests is the *relative efficiency* of two tests. Relative efficiency is defined as the *ratio of the sample sizes required by each test to obtain the same power value*. For example, suppose that for a given $\mu_1 - \mu_2$ value a t test had a power of .70 when $N = 10$ and a certain non-parametric test had the same power with an N of 12. According to our definition, the relative efficiency of the nonparametric test (compared to the t test) is $^{10}/_{12}$ or .833. This example illustrates the greater efficiency of the parametric test, the usual finding when sample N's are moderate or large in size. However, if very small samples are used (total N's of about 6), many nonparametric tests have relative efficiencies nearly equal to 1, compared to the corresponding parametric tests.

The relative efficiencies just mentioned were computed under the assumption that the data came from populations and samples satisfying the assumptions of the parametric test involved. Such data are precisely those for which parametric tests such as the t and the F test were developed. So nonparametric tests often work quite well even in cases where parametric tests are preferable. But once assumptions such as normality, which are required by the t and F tests, are known to be false, we know that t or F may lead to inappropriate conclusions about significance levels. In that case a nonparametric test will probably be preferable, even if it has low relative efficiency, because we can then be sure about our significance level.

Unfortunately, relative efficiency is not a constant that can be determined for any pair of tests. Rather, its value changes as the size of the sample or samples for one test changes. Relative efficiency also changes as other factors change, such as the α level desired and the value of the population difference (such as the value of $\mu_1 - \mu_2$) to be detected. Although we may want to know the relative efficiency for small-sample situations, it is more convenient to find a related measure, called the *asymptotic relative efficiency* (or A.R.E.), which is constant in value despite variations in such factors as we have mentioned.

What is asymptotic relative efficiency? *Asymptotic* means the final level reached or almost reached at the end or *asymptote* of a process, such as the level of proficiency reached on a motor task so well practiced that no further improvement is noticeable. Similarly, *asymptotic relative efficiency* is attained after N becomes large enough that relative efficiency does not change appreciably with further increases in the N for each group.

The asymptotic relative efficiencies for the five nonparametric tests described in this chapter are given in Table 15.6.

The tabled values help us decide what nonparametric test to use in a given situation. If we have two independent groups of scores with no tied ranks between groups, for example, we would prefer the Mann-Whitney U-test to the median test because of the latter's inefficiency, particularly with larger N's. Similarly, the Kruskal-Wallis test, which can be used with

Table 15.6 Asymptotic relative efficiencies (A.R.E.'s) of five nonparametric tests compared to the corresponding parametric tests.

Test	A.R.E.	Test	A.R.E.
Median test (two or more independent samples)	.637	Friedman rank test*	.637 for $k = 2$ .716 for $k = 3$
Mann-Whitney U-test (two independent samples)	.955		. . . .910 for $k = 20$
Kruskal-Wallis H-test (two or more independent samples)	.955		. . . .955 for $k = \infty$
Wilcoxon signed-rank test	.955		

* Where k = number of treatments given to each of N subjects.

more than two groups, is more efficient than the median test and should be used in preference to the latter. However, if there are several cases of tied ranks across groups, complications from these ties may make the median test preferable to the Kruskal-Wallis or Mann-Whitney, even though its A.R.E. is lower.

Nonparametric, distribution-free tests are a relatively recent development in statistics, and new techniques of this type continue to be devised to handle different kinds of statistical problems presented by empirical data. Although parametric tests continue to be preferable if the assumptions underlying their use can be met, nonparametric tests, a selected group of which we have presented in this chapter, are often of invaluable assistance to the research worker.

TERMS AND SYMBOLS TO REVIEW

Distribution-free test

Nonparametric vs. parametric test

Median test

Mann-Whitney U test

Kruskal-Wallis one-way analysis
of variance by ranks

Wilcoxon signed-ranks test

Friedman two-way analysis of
variance by ranks

Relative efficiency

Asymptotic relative efficiency
(A.R.E.)

List
of Formulas

Appendix **I**

Formula		Page

Formula **Page**

(5.9) $z = \dfrac{X - \bar{X}}{\tilde{\sigma}}$ (based on sample statistics) 59

$z = \dfrac{X - \mu}{\sigma}$ (based on population values)

(5.10) $X = \bar{X} + z(\tilde{\sigma})$ 60

(6.1) $p(A) = \dfrac{n_A}{N}$ 69

(6.2) $p(A \text{ or } B) = p(A) + p(B)$ 70

(6.3) $p(A, B) = p(A) \times p(B)$ (assuming independence) 71

$p(A, B) = p(A) \times p(B \mid A)$ (general formula) see 75–76

(6.4) $p(A, B, C) = p(A) \times p(B \mid A) \times p(C \mid A, B)$ 73

(6.5) $p(A \mid B) = \dfrac{p(A, B)}{p(B)}$ 75

(6.6) $p(B \mid A) = \dfrac{p(A, B)}{p(A)}$ 76

(6.7) $P_N = N! = N(N - 1)(N - 2) \ldots (1)$ 78

(6.8) $P_r^N = \dfrac{N!}{(N - r)!}$ 79

(6.9) $C_r^N = \dfrac{N!}{r!(N - r)!}$ 79

(7.1) $(p + q)^N = p^N + Np^{N-1}q + \dfrac{N(N - 1)}{1 \times 2} p^{N-2}q^2$

$+ \dfrac{N(N - 1)(N - 2)}{1 \times 2 \times 3} p^{N-3}q^3 + \cdots + q^N$ 84

(7.2) $C_r^N p^r q^{N-r} = \dfrac{N!}{r!(N - r)!} p^r q^{N-r}$ 85

(7.3) $\mu_b = Np$ 86

(7.4) $\sigma_b^2 = Npq$ 86

(7.5) $\sigma_b = \sqrt{Npq}$ 86

Formula	**Page**

(8.1) $\quad \sigma_{\bar{X}} = \dfrac{\sigma}{\sqrt{N}}$

$\hspace{9cm}$ 103

(8.2) $\quad z = \dfrac{\bar{X} - \mu}{\sigma_{\bar{X}}} = \dfrac{x}{\sigma_{\bar{X}}}$

$\hspace{9cm}$ 103

(8.3) $\quad s_{\bar{X}} = \dfrac{\tilde{\sigma}}{\sqrt{N - 1}}$

$\hspace{9cm}$ 104

(8.4) $\quad s_{\bar{X}} = \dfrac{s}{\sqrt{N}}$

$\hspace{9cm}$ 104

(8.5) $\quad t = \dfrac{\bar{X} - \mu}{s_{\bar{X}}} \qquad$ where $df = N - 1$ $\hspace{2cm}$ 106

(8.6) $\quad$ 99% CI for $\mu = \bar{X} \pm t_{(.01)}(s_{\bar{X}})$ $\hspace{3cm}$ 114

(8.7) $\quad$ 95% CI for $\mu = \bar{X} \pm t_{(.05)}(s_{\bar{X}})$ $\hspace{3cm}$ 115

(8.8) $\quad \sigma_{\text{prop}} = \sqrt{\dfrac{p_t q_t}{N}}$

$\hspace{9cm}$ 119

(8.9) $\quad z = \dfrac{p_s - p_t}{\sigma_{\text{prop}}}$

$\hspace{9cm}$ 120

(9.1) $\quad \sigma_{\bar{X}_1 - \bar{X}_2} = \sqrt{\sigma_{\bar{X}_1}{}^2 + \sigma_{\bar{X}_2}{}^2}$ $\hspace{4cm}$ 124

(9.2) $\quad z = \dfrac{(\bar{X}_1 - \bar{X}_2) - (\mu_1 - \mu_2)}{\sigma_{\bar{X}_1 - \bar{X}_2}}$ $\hspace{3.5cm}$ 124

(9.3) $\quad s_{\bar{X}_1 - \bar{X}_2} = \sqrt{s_{\bar{X}_1}{}^2 + s_{\bar{X}_2}{}^2} \qquad$ (equal N's) $\hspace{1.5cm}$ 125

(9.4) $\quad t = \dfrac{(\bar{X}_1 - \bar{X}_2) - (\mu_1 - \mu_2)}{s_{\bar{X}_1 - \bar{X}_2}}$ $\hspace{3.5cm}$ 126

(9.5) $\quad t = \dfrac{\bar{X}_1 - \bar{X}_2}{s_{\bar{X}_1 - \bar{X}_2}}$

$\hspace{2cm}$ where $H_0 : \mu_1 - \mu_2 = 0 \quad$ and $\quad df = N_1 + N_2 - 2$ $\hspace{0.5cm}$ 127

(9.6) $\quad s_{\bar{X}_1 - \bar{X}_2} = \sqrt{\dfrac{N_1 \tilde{\sigma}_1{}^2 + N_2 \tilde{\sigma}_2{}^2}{(N_1 + N_2 - 2)} \left(\dfrac{1}{N_1} + \dfrac{1}{N_2} \right)} \qquad$ (unequal N's)

$\hspace{2cm} = \sqrt{\dfrac{(N_1 - 1)s_1{}^2 + (N_2 - 1)s_2{}^2}{(N_1 + N_2 - 2)} \left(\dfrac{1}{N_1} + \dfrac{1}{N_2} \right)}$ $\hspace{1.5cm}$ 128

Formula **Page**

(9.7) $s_{\overline{X}_1 - \overline{X}_2} =$

$$\sqrt{\frac{(\sum X_1^2 + \sum X_2^2) - (N_1\overline{X}_1^2 + N_2\overline{X}_2^2)}{(N_1 + N_2 - 2)}\left(\frac{1}{N_1} + \frac{1}{N_2}\right)} \quad 129$$

(9.8) $df = N_1 - 1 + N_2 - 1 = N_1 + N_2 - 2$ (use with Eq. 9.5) 129

(9.9) Power $= 1 - p$ (Type II error) $= 1 - \beta$ 135

(10.1) $r = \dfrac{\sum z_X z_Y}{N}$ where $df = N - 2$

$$z_X = \frac{X - \overline{X}}{\tilde{\sigma}_X} \quad \text{and} \quad z_Y = \frac{Y - \overline{Y}}{\tilde{\sigma}_Y} \quad 146$$

(10.2) $r = \dfrac{\sum XY - N\overline{X}\overline{Y}}{\sqrt{\sum X^2 - N\overline{X}^2}\sqrt{\sum Y^2 - N\overline{Y}^2}}$ 148

(10.3) $Y_{pred} = \left(\dfrac{rs_Y}{s_X}\right)X - \left(\dfrac{rs_Y}{s_X}\right)\overline{X} + \overline{Y}$ 154

(10.4) $X_{pred} = \left(\dfrac{rs_X}{s_Y}\right)Y - \left(\dfrac{rs_X}{s_Y}\right)\overline{Y} + \overline{X}$ 155

(10.5) $r_S = 1 - \dfrac{6\sum d^2}{N(N^2 - 1)}$ 157

(11.1) $s_{\overline{X}_1 - \overline{X}_2} = \sqrt{s_{\overline{X}_1}^2 + s_{\overline{X}_2}^2 - 2r_{12}s_{\overline{X}_1}s_{\overline{X}_2}}$ (matched pairs) 168

(11.2) $\tilde{\sigma}_D^2 = \dfrac{\sum D^2}{N} - \overline{X}_D^2$ 172

(11.3) $s_{\overline{X}_D} = \sqrt{\dfrac{\tilde{\sigma}_D^2}{N - 1}}$ 172

(11.4) $t = \dfrac{\overline{X}_D}{s_{\overline{X}_D}} = \dfrac{\overline{X}_1 - \overline{X}_2}{\sqrt{s_{\overline{X}_1}^2 + s_{\overline{X}_2}^2 - 2r_{12}s_{\overline{X}_1}s_{\overline{X}_2}}}$

where $df = N - 1$ 172

(12.1) $SS_{tot} = \sum(X - \overline{X}_{tot})^2$ 181

(12.2) $SS_{wg} = SS_{wg1} + SS_{wg2} + S_{wg3} + \cdots + SS_{wgk}$ 182

(12.3) $SS_{bg} = \sum N_g(\overline{X}_g - \overline{X}_{tot})^2$ 183

Formula **Page**

(12.4) $\quad SS_{tot} = \left(\displaystyle\sum_{tot} X^2\right) - \dfrac{\left(\sum_{tot} X\right)^2}{N_{tot}}$ 185

(12.5) $\quad SS_{bg} = \displaystyle\sum_{g}\left[\dfrac{(\sum X_g)^2}{N_g}\right] - \dfrac{\left(\sum_{tot} X\right)^2}{N_{tot}}$ 185

(12.6) $\quad SS_{wg} = SS_{tot} - SS_{bg}$ 186

(12.7) $\quad SS_{wg} = \displaystyle\sum_{g}\left[(\textstyle\sum X_g^2) - \dfrac{(\sum X_g)^2}{N_g}\right]$ 186

(12.8) $\quad MS_{bg} = \dfrac{SS_{bg}}{df_{bg}} \quad$ where $df_{bg} = k - 1$ 188

(12.9) $\quad MS_{wg} = \dfrac{SS_{wg}}{df_{wg}} \quad$ where $df_{wg} = N_{tot} - k$ 188

(12.10) $\quad F = \dfrac{MS_{bg}}{MS_{wg}}$ 189

(12.11) $\quad hsd = q_\alpha \sqrt{\dfrac{MS_{wg}}{N_g}}$ 196

(13.1) $\quad SS_{tot} = \left(\displaystyle\sum_{tot} X^2\right) - \dfrac{\left(\sum_{tot} X\right)^2}{N_{tot}}$ 208

(13.2) $\quad SS_{bg} = \displaystyle\sum_{g}\left[\dfrac{(\sum X_g)^2}{N_g}\right] - \dfrac{\left(\sum_{tot} X\right)^2}{N_{tot}}$ 208

(13.3) $\quad SS_{wg} = SS_{tot} - SS_{bg}$ 208

(13.4) $\quad SS_{wg} = \displaystyle\sum_{g}\left[(\textstyle\sum X_g^2) - \dfrac{(\sum X_g)^2}{N_g}\right]$ 208

(13.5) $\quad SS_A = \dfrac{(\sum X_{AI})^2}{N_{AI}} + \dfrac{(\sum X_{AII})^2}{N_{AII}} + \cdots + \dfrac{(\sum X_{Am})^2}{N_{Am}} - \dfrac{\left(\sum_{tot} X\right)^2}{N_{tot}}$ 208

(13.6) $\quad SS_B = \dfrac{(\sum X_{BI})^2}{N_{BI}} + \dfrac{(\sum X_{BII})^2}{N_{BII}} + \cdots + \dfrac{(\sum X_{Bn})^2}{N_{Bn}} - \dfrac{\left(\sum_{tot} X\right)^2}{N_{tot}}$ 209

(13.7) $\quad SS_{A \times B} = SS_{bg} - SS_A - SS_B$ 209

Formula

(13.8) $\quad SS_{A \times B} = N \left[\sum_{g} (_{AB} - \overline{X}_A - \overline{X}_B + \overline{X}_{tot})^2 \right]$

(13.9) $\quad MS_A = \dfrac{SS_A}{df_A} \qquad\qquad M_{A \times B} = \dfrac{SS_{A \times B}}{df_{A \times B}}$

$\quad$ where $df_A = m - 1$; $\qquad$ where $df_{A \times B} = (m - 1)(n - 1)$

$\quad MS_B = \dfrac{SS_B}{df_B} \qquad\qquad MS_{wg} = \dfrac{SS_{wg}}{df_{wg}}$

$\quad$ where $df_B = n - 1$; $\qquad$ where $df_{wg} = N_{tot} - (m)(n)$

(13.10) $\quad F_A = \dfrac{MS_A}{MS_{wg}}; \quad F_B = \dfrac{MS_B}{MS_{wg}}; \quad F_{A \times B} = \dfrac{MS_{A \times B}}{MS_{wg}}$

(14.1) $\quad \chi^2 = \sum \dfrac{(O - E)^2}{E} \quad$ where $df = (\text{col.} - 1)(\text{row} - 1)$

(14.2) $\quad \chi^2 = \dfrac{N(AD - BC)^2}{(A + B)(C + D)(A + C)(B + D)}$

(14.3) $\quad \dfrac{x}{\sigma} = \sqrt{2\chi^2} - \sqrt{2n - 1}$

(14.4) $\quad \chi^2 = \dfrac{N(|AD - BC| - N/2)^2}{(A + B)(C + D)(A + C)(B + D)}$

(15.1) $\quad U_1 = N_1 N_2 + \dfrac{N_1(N_1 + 1)}{2} - R_1 \quad$ (Mann-Whitney U)

(15.2) $\quad U_E = \dfrac{N_1 N_2}{2}$

(15.3) $\quad \sigma_U = \sqrt{\dfrac{N_1 N_2 (N_1 + N_2 + 1)}{12}}$

(15.4) $\quad z = \dfrac{U_1 - U_E}{\sigma_U}$

(15.5) $\quad H = \dfrac{12}{N(N + 1)} \sum \left(\dfrac{R_i^2}{n_i} \right) - 3(N + 1)$

$\qquad\qquad$ (Kruskal-Wallis one-way analysis of variance)

Formula **Page**

(15.6) $T_E = \dfrac{N(N + 1)}{4}$ (Wilcoxon signed-ranks) 240

(15.7) $\sigma_T = \sqrt{\dfrac{N(N + 1)(2N + 1)}{24}}$ 240

(15.8) $z = \dfrac{T - T_E}{\sigma_T}$ 240

(15.9) $\chi_R^2 = \dfrac{12}{Nk(k + 1)} (\sum R_i^2) - 3N(k + 1)$

(Friedman two-way analysis of variance) 242

List
of Tables

Appendix II

Table A: Squares and Square Roots

Table B: Percent of Total Area Under the
Normal Curve

Table C: Critical Values of t

Table D: Values of r at the 5% and 1% Levels

Table E: Values of r_s at the 5% and 1% Levels

Table F: Values of F at the 5% and 1% Levels

Table G: Values of χ^2 at the 5% and 1% Levels

Table H: Values of T at the 5% and 1% Levels
in the Wilcoxon Signed-Ranks Test

Table I: Values of q_α (the Studentized Range
Statistic) at the 5% and 1% Levels

Table A* Table of squares and square roots of the numbers from 1 to 1000.

Number	Square	Square Root	Number	Square	Square Root
1	1	1.000	41	16 81	6.403
2	4	1.414	42	17 64	6.481
3	9	1.732	43	18 49	6.557
4	16	2.000	44	19 36	6.633
5	25	2.236	45	20 25	6.708
6	36	2.449	46	21 16	6.782
7	49	2.646	47	22 09	6.856
8	64	2.828	48	23 04	6.928
9	81	3.000	49	24 01	7.000
10	1 00	3.162	50	25 00	7.071
11	1 21	3.317	51	26 01	7.141
12	1 44	3.464	52	27 04	7.211
13	1 69	3.606	53	28 09	7.280
14	1 96	3.742	54	29 16	7.348
15	2 25	3.873	55	30 25	7.416
16	2 56	4.000	56	31 36	7.483
17	2 89	4.123	57	32 49	7.550
18	3 24	4.243	58	33 64	7.616
19	3 61	4.359	59	34 81	7.681
20	4 00	4.472	60	36 00	7.746
21	4 41	4.583	61	37 21	7.810
22	4 84	4.690	62	38 44	7.874
23	5 29	4.796	63	39 69	7.937
24	5 76	4.899	64	40 96	8.000
25	6 25	5.000	65	42 25	8.062
26	6 76	5.099	66	43 56	8.124
27	7 29	5.196	67	44 89	8.185
28	7 84	5.292	68	46 24	8.246
29	8 41	5.385	69	47 61	8.307
30	9 00	5.477	70	49 00	8.367
31	9 61	5.568	71	50 41	8.426
32	10 24	5.657	72	51 84	8.485
33	10 89	5.745	73	53 29	8.544
34	11 56	5.831	74	54 76	8.602
35	12 25	5.916	75	56 25	8.660
36	12 96	6.000	76	57 76	8.718
37	13 69	6.083	77	59 29	8.775
38	14 44	6.164	78	60 84	8.832
39	15 21	6.245	79	62 41	8.888
40	16 00	6.325	80	64 00	8.944

* Table A is reprinted from Table II of Lindquist, E. L., *A First Course in Statistics* (revised edition), published by Houghton Mifflin Company, by permission of the publishers.

Table of squares and square roots, *continued.*

Number	Square	Square Root	Number	Square	Square Root
81	65 61	9.000	121	1 46 41	11.000
82	67 24	9.055	122	1 48 84	11.045
83	68 89	9.110	123	1 51 29	11.091
84	70 56	9.165	124	1 53 76	11.136
85	72 25	9.220	125	1 56 25	11.180
86	73 96	9.274	126	1 58 76	11.225
87	75 69	9.327	127	1 61 29	11.269
88	77 44	9.381	128	1 63 84	11.314
89	79 21	9.434	129	1 66 41	11.358
90	81 00	9.487	130	1 69 00	11.402
91	82 81	9.539	131	1 71 61	11.446
92	84 64	9.592	132	1 74 24	11.489
93	86 49	9.644	133	1 76 89	11.533
94	88 36	9.695	134	1 79 56	11.576
95	90 25	9.747	135	1 82 25	11.619
96	92 16	9.798	136	1 84 96	11.662
97	94 09	9.849	137	1 87 69	11.705
98	96 04	9.899	138	1 90 44	11.747
99	98 01	9.950	139	1 93 21	11.790
100	1 00 00	10.000	140	1 96 00	11.832
101	1 02 01	10.050	141	1 98 81	11.874
102	1 04 04	10.100	142	2 01 64	11.916
103	1 06 09	10.149	143	2 04 49	11.958
104	1 08 16	10.198	144	2 07 36	12.000
105	1 10 25	10.247	145	2 10 25	12.042
106	1 12 36	10.296	146	2 13 16	12.083
107	1 14 49	10.344	147	2 16 09	12.124
108	1 16 64	10.392	148	2 19 04	12.166
109	1 18 81	10.440	149	2 22 01	12.207
110	1 21 00	10.488	150	2 25 00	12.247
111	1 23 21	10.536	151	2 28 01	12.288
112	1 25 44	10.583	152	2 31 04	12.329
113	1 27 69	10.630	153	2 34 09	12.369
114	1 29 96	10.677	154	2 37 16	12.410
115	1 32 25	10.724	155	2 40 25	12.450
116	1 34 56	10.770	156	2 43 36	12.490
117	1 36 89	10.817	157	2 46 49	12.530
118	1 39 24	10.863	158	2 49 64	12.570
119	1 41 61	10.909	159	2 52 81	12.610
120	1 44 00	10.954	160	2 56 00	12.649

Table of squares and square roots, *continued.*

Number	Square	Square Root	Number	Square	Square Root
161	2 59 21	12.689	201	4 04 01	14.177
162	2 62 44	12.728	202	4 08 04	14.213
163	2 65 69	12.767	203	4 12 09	14.248
164	2 68 96	12.806	204	4 16 16	14.283
165	2 72 25	12.845	205	4 20 25	14.318
166	2 75 56	12.884	206	4 24 36	14.353
167	2 78 89	12.923	207	4 28 49	14.387
168	2 82 24	12.961	208	4 32 64	14.422
169	2 85 61	13.000	209	4 36 81	14.457
170	2 89 00	13.038	210	4 41 00	14.491
171	2 92 41	13.077	211	4 45 21	14.526
172	2 95 84	13.115	212	4 49 44	14.560
173	2 99 29	13.153	213	4 53 69	14.595
174	3 02 76	13.191	214	4 57 96	14.629
175	3 06 25	13.229	215	4 62 25	14.663
176	3 09 76	13.266	216	4 66 56	14.697
177	3 13 29	13.304	217	4 70 89	14.731
178	3 16 84	13.342	218	5 75 24	14.765
179	3 20 41	13.379	219	4 79 61	14.799
180	3 24 00	13.416	220	4 84 00	14.832
181	3 27 61	13.454	221	4 88 41	14.866
182	3 31 24	13.491	222	4 92 84	14.900
183	3 34 89	13.528	223	4 97 29	14.933
184	3 38 56	13.565	224	5 01 76	14.967
185	3 42 25	13.601	225	5 06 25	15.000
186	3 45 96	13.638	226	5 10 76	15.033
187	3 49 69	13.675	227	5 15 29	15.067
188	3 53 44	13.711	228	5 19 84	15.100
189	3 57 21	13.748	229	5 24 41	15.133
190	3 61 00	13.784	230	5 29 00	15.166
191	3 64 81	13.820	231	5 33 61	15.199
192	3 68 64	13.856	232	5 38 24	15.232
193	3 72 49	13.892	233	5 42 89	15.264
194	3 76 36	13.928	234	5 47 56	15.297
195	3 80 25	13.964	235	5 52 25	15.330
196	3 84 16	14.000	236	5 56 96	15.362
197	3 88 09	14.036	237	5 61 69	15.395
198	3 92 04	14.071	238	5 66 44	15.427
199	3 96 01	14.107	239	5 71 21	15.460
200	4 00 00	14.142	240	5 76 00	15.492

Table of squares and square roots, *continued.*

Number	Square	Square Root	Number	Square	Square Root
241	5 80 81	15.524	281	7 89 61	16.763
242	5 85 64	15.556	282	7 95 24	16.793
243	5 90 49	15.588	283	8 00 89	16.823
244	5 95 36	15.620	284	8 06 56	16.852
245	6 00 25	15.652	285	8 12 25	16.882
246	6 05 16	15.684	286	8 17 96	16.912
247	6 10 09	15.716	287	8 23 69	16.941
248	6 15 04	15.748	288	8 29 44	16.971
249	6 20 01	15.780	289	8 35 21	17.000
250	6 25 00	15.811	290	8 41 00	17.029
251	6 30 01	15.843	291	8 46 81	17.059
252	6 35 04	15.875	292	8 52 64	17.088
253	6 40 09	15.906	293	8 58 49	17.117
254	6 45 16	15.937	294	8 64 36	17.146
255	6 50 25	15.969	295	8 70 25	17.176
256	6 55 36	16.000	296	8 76 16	17.205
257	6 60 49	16.031	297	8 82 09	17.234
258	6 65 64	16.062	298	8 88 04	17.263
259	6 70 81	16.093	299	8 94 01	17.292
260	6 76 00	16.125	300	9 00 00	17.321
261	6 81 21	16.155	301	9 06 01	17.349
262	6 86 44	16.186	302	9 12 04	17.378
263	6 91 69	16.217	303	9 18 09	17.407
264	6 96 96	16.248	304	9 24 16	17.436
265	7 02 25	16.279	305	9 30 25	17.464
266	7 07 56	16.310	306	9 36 36	17.493
267	7 12 89	16.340	307	9 42 49	17.521
268	7 18 24	16.371	308	9 48 64	17.550
269	7 23 61	16.401	309	9 54 81	17.578
270	7 29 00	16.432	310	9 61 00	17.607
271	7 34 41	16.462	311	9 67 21	17.635
272	7 39 84	16.492	312	9 73 44	17.664
273	7 45 29	16.523	313	9 79 69	17.692
274	7 50 76	16.553	314	9 85 96	17.720
275	7 56 25	16.583	315	9 92 25	17.748
276	7 61 76	16.613	316	9 98 56	17.776
277	7 67 29	16.643	317	10 04 89	17.804
278	7 72 84	16.673	318	10 11 24	17.833
279	7 78 41	16.703	319	10 17 61	17.861
280	7 84 00	16.733	320	10 24 00	17.889

Table of squares and square roots, *continued.*

Number	Square	Square Root	Number	Square	Square Root
321	10 30 41	17.916	361	13 03 21	19.000
322	10 36 84	17.944	362	13 10 44	19.026
323	10 43 29	17.972	363	13 17 69	19.053
324	10 49 76	18.000	364	13 24 96	19.079
325	10 56 25	18.028	365	13 32 25	19.105
326	10 62 76	18.055	366	13 39 56	19.131
327	10 69 29	18.083	367	13 46 89	19.157
328	10 75 84	18.111	368	13 54 24	19.183
329	10 82 41	18.138	369	13 61 61	19.209
330	10 89 00	18.166	370	13 69 00	19.235
331	10 95 61	18.193	371	13 76 41	19.261
332	11 02 24	18.221	372	13 83 84	19.287
333	11 08 89	18.248	373	13 91 29	19.313
334	11 15 56	18.276	374	13 98 76	19.339
335	11 22 25	18.303	375	14 06 25	19.363
336	11 28 96	18.330	376	14 13 76	19.391
337	11 35 69	18.358	377	14 21 29	19.416
338	11 42 44	18.385	378	14 28 84	19.442
339	11 49 21	18.412	379	14 36 41	19.468
340	11 56 00	18.439	380	14 44 00	19.494
341	11 62 81	18.466	381	14 51 61	19.519
342	11 69 64	18.493	382	14 59 24	19.545
343	11 76 49	18.520	383	14 66 89	19.570
344	11 83 36	18.547	384	14 74 56	19.596
345	11 90 25	18.574	385	14 82 25	19.621
346	11 97 16	18.601	386	14 89 96	19.647
347	12 04 09	18.628	387	14 97 69	19.672
348	12 11 04	18.655	388	15 05 44	19.698
349	12 18 01	18.682	389	15 13 21	19.723
350	12 25 00	18.708	390	15 21 00	19.748
351	12 32 01	18.735	391	15 28 81	19.774
352	12 39 04	18.762	392	15 36 64	19.799
353	12 46 09	18.788	393	15 44 49	19.824
354	12 53 16	18.815	394	15 52 36	19.849
355	12 60 25	18.841	395	15 60 25	19.875
356	12 67 36	18.868	396	15 68 16	19.900
357	12 74 49	18.894	397	15 76 09	19.925
358	12 81 64	18.921	398	15 84 04	19.950
359	12 88 81	18.947	399	15 92 01	19.975
360	12 96 00	18.974	400	16 00 00	20.000

Table of squares and square roots, *continued.*

Number	Square	Square Root	Number	Square	Square Root
401	16 08 01	20.025	441	19 44 81	21.000
402	16 16 04	20.050	442	19 53 64	21.024
403	16 24 09	20.075	443	19 62 49	21.048
404	16 32 16	20.100	444	19 71 36	21.071
405	16 40 25	20.125	445	19 80 25	21.095
406	16 48 36	20.149	446	19 89 16	21.119
407	16 56 49	20.174	447	19 98 09	21.142
408	16 64 64	20.199	448	20 07 04	21.166
409	16 72 81	20.224	449	20 16 01	21.190
410	16 81 00	20.248	450	20 25 00	21.213
411	16 89 21	20.273	451	20 34 01	21.237
412	16 97 44	20.298	452	20 43 04	21.260
413	17 05 69	20.322	453	20 52 09	21.284
414	17 13 96	20.347	454	20 61 16	21.307
415	17 22 25	20.372	455	20 70 25	21.331
416	17 30 56	20.396	456	20 79 36	21.354
417	17 38 89	20.421	457	20 88 49	21.378
418	17 47 24	20.445	458	20 97 64	21.401
419	17 55 61	20.469	459	21 06 81	21.424
420	17 64 00	20.494	460	21 16 00	21.448
421	17 72 41	20.518	461	21 25 21	21.471
422	17 80 84	20.543	462	21 34 44	21.494
423	17 89 29	20.567	463	21 43 69	21.517
424	17 97 76	20.591	464	21 52 96	21.541
425	18 06 25	20.616	465	21 62 25	21.564
426	18 14 76	20.640	466	21 71 56	21.587
427	18 23 29	20.664	467	21 80 89	21.610
428	18 31 84	20.688	468	21 90 24	21.633
429	18 40 41	20.712	469	21 99 61	21.656
430	18 49 00	20.736	470	22 09 00	21.679
431	18 57 61	20.761	471	22 18 41	21.703
432	18 66 24	20.785	472	22 27 84	21.726
433	18 74 89	20.809	473	22 37 29	21.749
434	18 83 56	20.833	474	22 46 76	21.772
435	18 92 25	20.857	475	22 56 25	21.794
436	19 00 96	20.881	476	22 65 76	21.817
437	19 09 69	20.905	477	22 75 29	21.840
438	19 18 44	20.928	478	22 84 84	21.863
439	19 27 21	20.952	479	22 94 41	21.886
440	19 36 00	20.976	480	23 04 00	21.909

Table of squares and square roots, *continued*.

Number	Square	Square Root	Number	Square	Square Root
481	23 13 61	21.932	521	27 14 41	22.825
482	23 23 24	21.954	522	27 24 84	22.847
483	23 32 89	21.977	523	27 35 29	22.869
484	23 42 56	22.000	524	27 45 76	22.891
485	23 52 25	22.023	525	27 56 25	22.913
486	23 61 96	22.045	526	27 66 76	22.935
487	23 71 69	22.068	527	27 77 29	22.956
488	23 81 44	22.091	528	27 87 84	22.978
489	23 91 21	22.113	529	27 98 41	23.000
490	24 01 00	22.136	530	28 09 00	23.022
491	24 10 81	22.159	531	28 19 61	23.043
492	24 20 64	22.181	532	28 30 24	23.065
493	24 30 49	22.204	533	28 40 89	23.087
494	24 40 36	22.226	534	28 51 56	23.108
495	24 50 25	22.249	535	28 62 25	23.130
496	24 60 16	22.271	536	28 72 96	23.152
497	24 70 09	22.293	537	28 83 69	23.173
498	24 80 04	22.316	538	28 94 44	23.195
499	24 90 01	22.338	539	29 05 21	23.216
500	25 00 00	22.361	540	29 16 00	23.238
501	25 10 01	22.383	541	29 26 81	23.259
502	25 20 04	22.405	542	29 37 64	23.281
503	25 30 09	22.428	543	29 48 49	23.302
504	25 40 16	22.450	544	29 59 36	23.324
505	25 50 25	22.472	545	29 70 25	23.345
506	25 60 36	22.494	546	29 81 16	23.367
507	25 70 49	22.517	547	29 92 09	23.388
508	25 80 64	22.539	548	30 03 04	23.409
509	25 90 81	22.561	549	30 14 01	23.431
510	26 01 00	22.583	550	30 25 00	23.452
511	26 11 21	22.605	551	30 36 01	23.473
512	26 21 44	22.627	552	30 47 04	23.495
513	26 31 69	22.650	553	30 58 09	23.516
514	26 41 96	22.672	554	30 69 16	23.537
515	26 52 25	22.694	555	30 80 25	23.558
516	26 62 56	22.716	556	30 91 36	23.580
517	26 72 89	22.738	557	31 02 49	23.601
518	26 83 24	22.760	558	31 13 64	23.622
519	26 93 61	22.782	559	31 24 81	23.643
520	27 04 00	22.804	560	31 36 00	23.664

Table of squares and square roots, *continued*.

Number	Square	Square Root	Number	Square	Square Root
561	31 47 21	23.685	601	36 12 01	24.515
562	31 58 44	23.707	602	36 24 04	24.536
563	31 69 69	23.728	603	36 36 09	24.556
564	31 80 96	23.749	604	36 48 16	24.576
565	31 92 25	23.770	605	36 60 25	24.597
566	32 03 56	23.791	606	36 72 36	24.617
567	32 14 89	23.812	607	36 84 49	24.637
568	32 26 24	23.833	608	36 96 64	24.658
569	32 37 61	23.854	609	37 08 81	24.678
570	32 49 00	23.875	610	37 21 00	24.698
571	32 60 41	23.896	611	37 33 21	24.718
572	32 71 84	23.917	612	37 45 44	24.739
573	32 83 29	23.937	613	37 57 69	24.759
574	32 94 76	23.958	614	37 69 96	24.779
575	33 06 25	23.979	615	37 82 25	24.799
576	33 17 76	24.000	616	37 94 56	24.819
577	33 29 29	24.021	617	38 06 89	24.839
578	33 40 84	24.042	618	38 19 24	24.860
579	33 52 41	24.062	619	38 31 61	24.880
580	33 64 00	24.083	620	38 44 00	24.900
581	33 75 61	24.104	621	38 56 41	24.920
582	33 87 24	24.125	622	38 68 84	24.940
583	33 98 89	24.145	623	38 81 29	24.960
584	34 10 56	24.166	624	38 93 76	24.980
585	34 22 25	24.187	625	39 06 25	25.000
586	34 33 96	24.207	626	39 18 76	25.020
587	34 45 69	24.228	627	39 31 29	25.040
588	34 57 44	24.249	628	39 43 84	25.060
589	34 69 21	24.269	629	39 56 41	25.080
590	34 81 00	24.290	630	39 69 00	25.100
591	34 92 81	24.310	631	39 81 61	25.120
592	35 04 64	24.331	632	39 94 24	25.140
593	35 16 49	24.352	633	40 06 89	25.159
594	35 28 36	24.372	634	40 19 56	25.179
595	35 40 25	24.393	635	40 32 25	25.199
596	35 52 16	24.413	636	40 44 96	25.219
597	35 64 09	24.434	637	40 57 69	25.239
598	35 76 04	24.454	638	40 70 44	25.259
599	35 88 01	24.474	639	40 83 21	25.278
600	36 00 00	24.495	640	40 96 00	25.298

Table of squares and square roots, *continued*.

Number	Square	Square Root	Number	Square	Square Root
641	41 08 81	25.318	681	46 37 61	26.096
642	41 21 64	25.338	682	46 51 24	26.115
643	41 34 49	25.357	683	46 64 89	26.134
644	41 47 36	25.377	684	46 78 56	26.153
645	41 60 25	25.397	685	46 92 25	26.173
646	41 73 16	25.417	686	47 05 96	26.192
647	41 86 09	25.436	687	47 19 69	26.211
648	41 99 04	25.456	688	47 33 44	26.230
649	42 12 01	25.475	689	47 47 21	26.249
650	42 25 00	25.495	690	47 61 00	26.268
651	42 38 01	25.515	691	47 74 81	26.287
652	42 51 04	25.534	692	47 88 64	26.306
653	42 64 09	25.554	693	48 02 49	26.325
654	42 77 16	25.573	694	48 16 36	26.344
655	42 90 25	25.593	695	48 30 25	26.363
656	43 03 36	25.612	696	48 44 16	26.382
657	43 16 49	25.632	697	48 58 09	26.401
658	43 29 64	25.652	698	48 72 04	26.420
659	43 42 81	25.671	699	48 86 01	26.439
660	43 56 00	25.690	700	49 00 00	26.458
661	43 69 21	25.710	701	49 14 01	26.476
662	43 82 44	25.729	702	49 28 04	26.495
663	43 95 69	25.749	703	49 42 09	26.514
664	44 08 96	25.768	704	49 56 16	26.533
665	44 22 25	25.788	705	49 70 25	26.552
666	44 35 56	25.807	706	49 84 36	26.571
667	44 48 89	25.826	707	49 98 49	26.589
668	44 62 24	25.846	708	50 12 64	26.608
669	44 75 61	25.865	709	50 26 81	26.627
670	44 89 00	25.884	710	50 41 00	26.646
671	45 02 41	25.904	711	50 55 21	26.665
672	45 15 84	25.923	712	50 69 44	26.683
673	45 29 29	25.942	713	50 83 69	26.702
674	45 42 76	25.962	714	50 97 96	26.721
675	45 56 25	25.981	715	51 12 25	26.739
676	45 69 76	26.000	716	51 26 56	26.758
677	45 83 29	26.019	717	51 40 89	26.777
678	45 96 84	26.038	718	51 55 24	26.796
679	46 10 41	26.058	719	51 69 61	26.814
680	46 24 00	26.077	720	51 84 00	26.833

Table of squares and square roots, *continued.*

Number	Square	Square Root	Number	Square	Square Root
721	51 98 41	26.851	761	57 91 21	27.586
722	52 12 84	26.870	762	58 06 44	27.604
723	52 27 29	26.889	763	58 21 69	27.622
724	52 41 76	26.907	764	58 36 96	27.641
725	52 56 25	26.926	765	58 52 25	27.659
726	52 70 76	26.944	766	58 67 56	27.677
727	52 85 29	26.963	767	58 82 89	27.695
728	52 99 84	26.981	768	58 98 24	27.713
729	53 14 41	27.000	769	59 13 61	27.731
730	53 29 00	27.019	770	59 29 00	27.749
731	53 43 61	27.037	771	59 44 41	27.767
732	53 58 24	27.055	772	59 59 84	27.785
733	53 72 89	27.074	773	59 75 29	27.803
734	53 87 56	27.092	774	59 90 76	27.821
735	54 02 25	27.111	775	60 06 25	27.839
736	54 16 96	27.129	776	60 21 76	27.857
737	54 31 69	27.148	777	60 37 29	27.875
738	54 46 44	27.166	778	60 52 84	27.893
739	54 61 21	27.185	779	60 68 41	27.911
740	54 76 00	27.203	780	60 84 00	27.928
741	54 90 81	27.221	781	60 99 61	27.946
742	55 05 64	27.240	782	61 15 24	27.964
743	55 20 49	27.258	783	61 30 89	27.982
744	55 35 36	27.276	784	61 46 56	28.000
745	55 50 25	27.295	785	61 62 25	28.018
746	55 65 16	27.313	786	61 77 96	28.036
747	55 80 09	27.331	787	61 93 69	28.054
748	55 95 04	27.350	788	62 09 44	28.071
749	56 10 01	27.368	789	62 25 21	28.089
750	56 25 00	27.386	790	62 41 00	28.107
751	56 40 01	27.404	791	62 56 81	28.125
752	56 55 04	27.423	792	62 72 64	28.142
753	56 70 09	27.441	793	62 88 49	28.160
754	56 85 16	27.459	794	63 04 36	28.178
755	57 00 25	27.477	795	63 20 25	28.196
756	57 15 36	27.495	796	63 36 16	28.213
757	57 30 49	27.514	797	63 52 09	28.231
758	57 45 64	27.532	798	63 68 04	28.249
759	57 60 81	27.550	799	63 84 01	28.267
760	57 76 00	27.568	800	64 00 00	28.284

Table of squares and square roots, *continued.*

Number	Square	Square Root	Number	Square	Square Root
801	64 16 01	28.302	841	70 72 81	29.000
802	64 32 04	28.320	842	70 89 64	29.017
803	64 48 09	28.337	843	71 06 49	29.034
804	64 64 16	28.355	844	71 23 36	29.052
805	64 80 25	28.373	845	71 40 25	29.069
806	64 96 36	28.390	846	71 57 16	29.086
807	65 12 49	28.408	847	71 74 09	29.103
808	65 28 64	28.425	848	71 91 04	29.120
809	65 44 81	28.443	849	72 08 01	29.138
810	65 61 00	28.460	850	72 25 00	29.155
811	65 77 21	28.478	851	72 42 01	29.172
812	65 93 44	28.496	852	72 59 04	29.189
813	66 09 69	28.513	853	72 76 09	29.206
814	66 25 96	28.531	854	72 93 16	29.223
815	66 42 25	28.548	855	73 10 25	29.240
816	66 58 56	28.566	856	73 27 36	29.257
817	66 74 89	28.583	857	73 44 49	29.275
818	66 91 24	28.601	858	73 61 64	29.292
819	67 07 61	28.618	859	73 78 81	29.309
820	67 24 00	28.636	860	73 96 00	29.326
821	67 40 41	28.653	861	74 13 21	29.343
822	67 56 84	28.671	862	74 30 44	29.360
823	67 73 29	28.688	863	74 47 69	29.377
824	67 89 76	28.705	864	74 64 96	29.394
825	68 06 25	28.723	865	74 82 25	29.411
826	68 22 76	28.740	866	74 99 56	29.428
827	68 39 29	28.758	867	75 16 89	29.445
828	68 55 84	28.775	868	75 34 24	29.462
829	68 72 41	28.792	869	75 51 61	29.479
830	68 89 00	28.810	870	75 69 00	29.496
831	69 05 61	28.827	871	75 86 41	29.513
832	69 22 24	28.844	872	76 03 84	29.530
833	69 38 89	28.862	873	76 21 29	29.547
834	69 55 56	28.879	874	76 38 76	29.563
835	69 72 25	28.896	875	76 56 25	29.580
836	69 88 96	28.914	876	76 73 76	29.597
837	70 05 69	28.931	877	76 91 29	29.614
838	70 22 44	28.948	878	77 08 84	29.631
839	70 39 21	28.965	879	77 26 41	29.648
840	70 56 00	28.983	880	77 44 00	29.665

Table of squares and square roots, *continued.*

Number	Square	Square Root	Number	Square	Square Root
881	77 61 61	29.682	921	84 82 41	30.348
882	77 79 24	29.698	922	85 00 84	30.364
883	77 96 89	29.715	923	85 19 29	30.381
884	78 14 56	29.732	924	85 37 76	30.397
885	78 32 25	29.749	925	85 56 25	30.414
886	78 49 96	29.766	926	85 74 76	30.430
887	78 67 69	29.783	927	85 93 29	30.447
888	78 85 44	29.799	928	86 11 84	30.463
889	79 03 21	29.816	929	86 30 41	30.480
890	79 21 00	29.833	930	86 49 00	30.496
891	79 38 81	29.850	931	86 67 61	30.512
892	79 56 64	29.866	932	86 86 24	30.529
893	79 74 49	29.883	933	87 04 89	30.545
894	79 92 36	29.900	934	87 23 56	30.561
895	80 10 25	29.916	935	87 42 25	30.578
896	80 28 16	29.933	936	87 60 96	30.594
897	80 46 09	29.950	937	87 79 69	30.610
898	80 64 04	29.967	938	87 98 44	30.627
899	80 82 01	29.983	939	88 17 21	30.643
900	81 00 00	30.000	940	88 36 00	30.659
901	81 18 01	30.017	941	88 54 81	30.676
902	81 36 04	30.033	942	88 73 64	30.692
903	81 54 09	30.050	943	88 92 49	30.708
904	81 72 16	30.067	944	89 11 36	30.725
905	81 90 25	30.083	945	89 30 25	30.741
906	82 08 36	30.100	946	89 49 16	30.757
907	82 26 49	30.116	947	89 68 09	30.773
908	82 44 64	30.133	948	89 87 04	30.790
909	82 62 81	30.150	949	90 06 01	30.806
910	82 81 00	30.166	950	90 25 00	30.822
911	82 99 21	30.183	951	90 44 01	30.838
912	83 17 44	30.199	952	90 63 04	30.854
913	83 35 69	30.216	953	90 82 09	30.871
914	83 53 96	30.232	954	91 01 16	30.887
915	83 72 25	30.249	955	91 20 25	30.903
916	83 90 56	30.265	956	91 39 36	30.919
917	84 08 89	30.282	957	91 58 49	30.935
918	84 27 24	30.299	958	91 77 64	30.952
919	84 45 61	30.315	959	91 96 81	30.968
920	84 64 00	30.332	960	92 16 00	30.984

Table of squares and square roots, *continued*.

Number	Square	Square Root	Number	Square	Square Root
961	92 35 21	31.000	981	96 23 61	31.321
962	92 54 44	31.016	982	96 43 24	31.337
963	92 73 69	31.032	983	96 62 89	31.353
964	92 92 96	31.048	984	96 82 56	31.369
965	93 12 25	31.064	985	97 02 25	31.385
966	93 31 56	31.081	986	97 21 96	31.401
967	93 50 89	31.097	987	97 41 69	31.417
968	93 70 24	31.113	988	97 61 44	31.432
969	93 89 61	31.129	989	97 81 21	31.448
970	94 09 00	31.145	990	98 01 00	31.464
971	94 28 41	31.161	991	98 20 81	31.480
972	94 47 84	31.177	992	98 40 64	31.496
973	94 67 29	31.193	993	98 60 49	31.512
974	94 86 76	31.209	994	98 80 36	31.528
975	95 06 25	31.225	995	99 00 25	31.544
976	95 25 76	31.241	996	99 20 16	31.559
977	95 45 29	31.257	997	99 40 09	31.575
978	95 64 84	31.273	998	99 60 04	31.591
979	95 84 41	31.289	999	99 80 01	31.607
980	96 04 00	31.305	1000	100 00 00	31.623

Table B* Percent of total area under the normal curve between mean ordinate and ordinate at any given sigma-distance from the mean.

$\dfrac{x}{\sigma}$	.00	.01	.02	.03	.04	.05	.06	.07	.08	.09
0.0	00.00	00.40	00.80	01.20	01.60	01.99	02.39	02.79	03.19	03.59
0.1	03.98	04.38	04.78	05.17	05.57	05.96	06.36	06.75	07.14	07.53
0.2	07.93	08.32	08.71	09.10	09.48	09.87	10.26	10.64	11.03	11.41
0.3	11.79	12.17	12.55	12.93	13.31	13.68	14.06	14.43	14.80	15.17
0.4	15.54	15.91	16.28	16.64	17.00	17.36	17.72	18.08	18.44	18.79
0.5	19.15	19.50	19.85	20.19	20.54	20.88	21.23	21.57	21.90	22.24
0.6	22.57	22.91	23.24	23.57	23.89	24.22	24.54	24.86	25.17	25.49
0.7	25.80	26.11	26.42	26.73	27.04	27.34	27.64	27.94	28.23	28.52
0.8	28.81	29.10	29.39	29.67	29.95	30.23	30.51	30.78	31.06	31.33
0.9	31.59	31.86	32.12	32.38	32.64	32.90	33.15	33.40	33.65	33.89
1.0	34.13	34.38	34.61	34.85	35.08	35.31	35.54	35.77	35.99	36.21
1.1	36.43	36.65	36.86	37.08	37.29	37.49	37.70	37.90	38.10	38.30
1.2	38.49	38.69	38.88	39.07	39.25	39.44	39.62	39.80	39.97	40.15
1.3	40.32	40.49	40.66	40.82	40.99	41.15	41.31	41.47	41.62	41.77
1.4	41.92	42.07	42.22	42.36	42.51	42.65	42.79	42.92	43.06	43.19
1.5	43.32	43.45	43.57	43.70	43.83	43.94	44.06	44.18	44.29	44.41
1.6	44.52	44.63	44.74	44.84	44.95	45.05	45.15	45.25	45.35	45.45
1.7	45.54	45.64	45.73	45.82	45.91	45.99	46.08	46.16	46.25	46.33
1.8	46.41	46.49	46.56	46.64	46.71	46.78	46.86	46.93	46.99	47.06
1.9	47.13	47.19	47.26	47.32	47.38	47.44	47.50	47.56	47.61	47.67
2.0	47.72	47.78	47.83	47.88	47.93	47.98	48.03	48.08	48.12	48.17
2.1	48.21	48.26	48.30	48.34	48.38	48.42	48.46	48.50	48.54	48.57
2.2	48.61	48.64	48.68	48.71	48.75	48.78	48.81	48.84	48.87	48.90
2.3	48.93	48.96	48.98	49.01	49.04	49.06	49.09	49.11	49.13	49.16
2.4	49.18	49.20	49.22	49.25	49.27	49.29	49.31	49.32	49.34	49.36
2.5	49.38	49.40	49.41	49.43	49.45	49.46	49.48	49.49	49.51	49.52
2.6	49.53	49.55	49.56	49.57	49.59	49.60	49.61	49.62	49.63	49.64
2.7	49.65	49.66	49.67	49.68	49.69	49.70	49.71	49.72	49.73	49.74
2.8	49.74	49.75	49.76	49.77	49.77	49.78	49.79	49.79	49.80	49.81
2.9	49.81	49.82	49.82	49.83	49.84	49.84	49.85	49.85	49.86	49.86
3.0	49.87									
3.5	49.98									
4.0	49.997									
5.0	49.99997									

* The original data for Table B came from *Tables for Statisticians and Biometricians*, edited by Karl Pearson, published by Cambridge University Press, and are used here by permission of the publisher. The adaptation of these data is taken from Lindquist, E. L., *A First Course in Statistics* (revised edition), with permission of the publisher, Houghton Mifflin Company.

Table C*   Critical values of *t*.

df	Level of significance for one-tailed test			
	5%	2.5%	1%	.5%
	Level of significance for two-tailed test			
	10%	5%	2%	1%
1	6.3138	12.7062	31.8207	63.6574
2	2.9200	4.3027	6.9646	9.9248
3	2.3534	3.1824	4.5407	5.8409
4	2.1318	2.7764	3.7469	4.6041
5	2.0150	2.5706	3.3649	4.0322
6	1.9432	2.4469	3.1427	3.7074
7	1.8946	2.3646	2.9980	3.4995
8	1.8595	2.3060	2.8965	3.3554
9	1.8331	2.2622	2.8214	3.2498
10	1.8125	2.2281	2.7638	3.1693
11	1.7959	2.2010	2.7181	3.1058
12	1.7823	2.1788	2.6810	3.0545
13	1.7709	2.1604	2.6503	3.0123
14	1.7613	2.1448	2.6245	2.9768
15	1.7531	2.1315	2.6025	2.9467
16	1.7459	2.1199	2.5835	2.9208
17	1.7396	2.1098	2.5669	2.8982
18	1.7341	2.1009	2.5524	2.8784
19	1.7291	2.0930	2.5395	2.8609
20	1.7247	2.0860	2.5280	2.8453
21	1.7207	2.0796	2.5177	2.8314
22	1.7171	2.0739	2.5083	2.8188
23	1.7139	2.0687	2.4999	2.8073
24	1.7109	2.0639	2.4922	2.7969
25	1.7081	2.0595	2.4851	2.7874
26	1.7056	2.0555	2.4786	2.7787
27	1.7033	2.0518	2.4727	2.7707
28	1.7011	2.0484	2.4671	2.7633
29	1.6991	2.0452	2.4620	2.7564
30	1.6973	2.0423	2.4573	2.7500

* Table C is abridged from Owen, D. B., *Handbook of Statistical Tables,* 1962, Addison-Wesley, Reading, Mass. Courtesy of the U.S. Energy Research and Development Administration.

Table of values of *t, continued.*

df	Level of significance for one-tailed test			
	5%	2.5%	1%	.5%
	Level of significance for two-tailed test			
df	10%	5%	2%	1%
35	1.6869	2.0301	2.4377	2.7238
40	1.6839	2.0211	2.4233	2.7045
45	1.6794	2.0141	2.4121	2.6896
50	1.6759	2.0086	2.4033	2.6778
60	1.6706	2.0003	2.3901	2.6603
70	1.6669	1.9944	2.3808	2.6479
80	1.6641	1.9901	2.3739	2.6387
90	1.6620	1.9867	2.3685	2.6316
100	1.6602	1.9840	2.3642	2.6259
110	1.6588	1.9818	2.3607	2.6213
120	1.6577	1.9799	2.3598	2.6174
∞	1.6449	1.9600	2.3263	2.5758

Table D* Values of *r* at the 5% and 1% levels of significance (two-tailed test).

Degrees of Freedom (*df*)	5%	1%	Degrees of Freedom (*df*)	5%	1%
1	.997	1.000	24	.388	.496
2	.950	.990	25	.381	.487
3	.878	.959	26	.374	.478
4	.811	.917	27	.367	.470
5	.754	.874	28	.361	.463
6	.707	.834	29	.355	.456
7	.666	.798	30	.349	.449
8	.632	.765	35	.325	.418
9	.602	.735	40	.304	.393
10	.576	.708	45	.288	.372
11	.553	.684	50	.273	.354
12	.532	.661	60	.250	.325
13	.514	.641	70	.232	.302
14	.497	.623	80	.217	.283
15	.482	.606	90	.205	.267
16	.468	.590	100	.195	.254
17	.456	.575	125	.174	.228
18	.444	.561	150	.159	.208
19	.433	.549	200	.138	.181
20	.423	.537	300	.113	.148
21	.413	.526	400	.098	.128
22	.404	.515	500	.088	.115
23	.396	.505	1000	.062	.081

* A portion of Table D is abridged from Table VI of Fisher and Yates, *Statistical Tables for Biological, Agricultural, and Medical Research,* 4th edition, 1953, published by Longman Group Ltd., London (previously published by Oliver and Boyd, Edinburgh), by permission of the authors and publishers. The remainder of the table is from Snedecor, *Statistical Methods,* by permission of the publisher, The Iowa State University Press, and the author.

Table E*   Values of r_s (rank-order correlation coefficient) at the 5% and 1% levels of significance (two-tailed test).

N	5%	1%
5	1.000	—
6	.886	1.000
7	.786	.929
8	.738	.881
9	.683	.833
10	.648	.794
12	.591	.777
14	.544	.714
16	.506	.665
18	.475	.625
20	.450	.591
22	.428	.562
24	.409	.537
26	.392	.515
28	.377	.496
30	.364	.478

* Computed from Olds, E. G., Distribution of the sum of squares of rank differences for small numbers of individuals, *Annals of Mathematical Statistics*, 1938, 9, 133–148, and, The 5% significance levels for sums of squares of rank differences and a correction, *Annals of Mathematical Statistics*, 1949, 20, 117–118, by permission of the author and the Institute of Mathematical Statistics.

Table F* Values of F at the 5% and 1% significance levels.

(*df associated with the denominator*)		\(df associated with the numerator\)								
		1	2	3	4	5	6	7	8	9
1	5%	161	200	216	225	230	234	237	239	241
	1%	**4052**	**5000**	**5403**	**5625**	**5764**	**5859**	**5928**	**5982**	**6022**
2	5%	18.5	19.0	19.2	19.2	19.3	19.3	19.4	19.4	19.4
	1%	**98.5**	**99.0**	**99.2**	**99.2**	**99.3**	**99.3**	**99.4**	**99.4**	**99.4**
3	5%	10.1	9.55	9.28	9.12	9.01	8.94	8.89	8.85	8.81
	1%	**34.1**	**30.8**	**29.5**	**28.7**	**28.2**	**27.9**	**27.7**	**27.5**	**27.3**
4	5%	7.71	6.94	6.59	6.39	6.26	6.16	6.09	6.04	6.00
	1%	**21.2**	**18.0**	**16.7**	**16.0**	**15.5**	**15.2**	**15.0**	**14.8**	**14.7**
5	5%	6.61	5.79	5.41	5.19	5.05	4.95	4.88	4.82	4.77
	1%	**16.3**	**13.3**	**12.1**	**11.4**	**11.0**	**10.7**	**10.5**	**10.3**	**10.2**
6	5%	5.99	5.14	4.76	4.53	4.39	4.28	4.21	4.15	4.10
	1%	**13.7**	**10.9**	**9.78**	**9.15**	**8.75**	**8.47**	**8.26**	**8.10**	**7.98**
7	5%	5.59	4.74	4.35	4.12	3.97	3.87	3.79	3.73	3.68
	1%	**12.2**	**9.55**	**8.45**	**7.85**	**7.46**	**7.19**	**6.99**	**6.84**	**6.72**
8	5%	5.32	4.46	4.07	3.84	3.69	3.58	3.50	3.44	3.39
	1%	**11.3**	**8.65**	**7.59**	**7.01**	**6.63**	**6.37**	**6.18**	**6.03**	**5.91**
9	5%	5.12	4.26	3.86	3.63	3.48	3.37	3.29	3.23	3.18
	1%	**10.6**	**8.02**	**6.99**	**6.42**	**6.06**	**5.80**	**5.61**	**5.47**	**5.35**
10	5%	4.96	4.10	3.71	3.48	3.33	3.22	3.14	3.07	3.02
	1%	**10.0**	**7.56**	**6.55**	**5.99**	**5.64**	**5.39**	**5.20**	**5.06**	**4.94**
11	5%	4.84	3.98	3.59	3.36	3.20	3.09	3.01	2.95	2.90
	1%	**9.65**	**7.21**	**6.22**	**5.67**	**5.32**	**5.07**	**4.89**	**4.74**	**4.63**
12	5%	4.75	3.89	3.49	3.26	3.11	3.00	2.91	2.85	2.80
	1%	**9.33**	**6.93**	**5.95**	**5.41**	**5.06**	**4.82**	**4.64**	**4.50**	**4.39**
13	5%	4.67	3.81	3.41	3.18	3.03	2.92	2.83	2.77	2.71
	1%	**9.07**	**6.70**	**5.74**	**5.21**	**4.86**	**4.62**	**4.44**	**4.30**	**4.19**
14	5%	4.60	3.74	3.34	3.11	2.96	2.85	2.76	2.70	2.65
	1%	**8.86**	**6.51**	**5.56**	**5.04**	**4.70**	**4.46**	**4.28**	**4.14**	**4.03**
15	5%	4.54	3.68	3.29	3.06	2.90	2.79	2.71	2.64	2.59
	1%	**8.68**	**6.36**	**5.42**	**4.89**	**4.56**	**4.32**	**4.14**	**4.00**	**3.89**
16	5%	4.49	3.63	3.24	3.01	2.85	2.74	2.66	2.59	2.54
	1%	**8.53**	**6.23**	**5.29**	**4.77**	**4.44**	**4.20**	**4.03**	**3.89**	**3.78**
17	5%	4.45	3.59	3.20	2.96	2.81	2.70	2.61	2.55	2.49
	1%	**8.40**	**6.11**	**5.18**	**4.67**	**4.34**	**4.10**	**3.93**	**3.79**	**3.68**

* Merrington, M., and Thompson, C. M. Tables of percentage points of the inverted beta (F) distribution, *Biometrika*, 1943, 33, 73–88, by permission of the editor.

Values of F at the 5% and 1% significance levels, *continued.*

(*df associated with the denominator*)		(*df associated with the numerator*)								
		1	2	3	4	5	6	7	8	9
18	5%	4.41	3.55	3.16	2.93	2.77	2.66	2.58	2.51	2.46
	1%	**8.29**	**6.01**	**5.09**	**4.58**	**4.25**	**4.01**	**3.84**	**3.71**	**3.60**
19	5%	4.38	3.52	3.13	2.90	2.74	2.63	2.54	2.48	2.42
	1%	**8.18**	**5.93**	**5.01**	**4.50**	**4.17**	**3.94**	**3.77**	**3.63**	**3.52**
20	5%	4.35	3.49	3.10	2.87	2.71	2.60	2.51	2.45	2.39
	1%	**8.10**	**5.85**	**4.94**	**4.43**	**4.10**	**3.87**	**3.70**	**3.56**	**3.46**
21	5%	4.32	3.47	3.07	2.84	2.68	2.57	2.49	2.42	2.37
	1%	**8.02**	**5.78**	**4.87**	**4.37**	**4.04**	**3.81**	**3.64**	**3.51**	**3.40**
22	5%	4.30	3.44	3.05	2.82	2.66	2.55	2.46	2.40	2.34
	1%	**7.95**	**5.72**	**4.82**	**4.31**	**3.99**	**3.76**	**3.59**	**3.45**	**3.35**
23	5%	4.28	3.42	3.03	2.80	2.64	2.53	2.44	2.37	2.32
	1%	**7.88**	**5.66**	**4.76**	**4.26**	**3.94**	**3.71**	**3.54**	**3.41**	**3.30**
24	·5%	4.26	3.40	3.01	2.78	2.62	2.51	2.42	2.36	2.30
	1%	**7.82**	**5.61**	**4.72**	**4.22**	**3.90**	**3.67**	**3.50**	**3.36**	**3.26**
25	5%	4.24	3.39	2.29	2.76	2.60	2.49	2.40	2.34	2.28
	1%	**7.77**	**5.57**	**4.68**	**4.18**	**3.86**	**3.63**	**3.46**	**3.32**	**3.22**
26	5%	4.23	3.37	2.98	2.74	2.59	2.47	2.39	2.32	2.27
	1%	**7.72**	**5.53**	**4.64**	**4.14**	**3.82**	**3.59**	**3.42**	**3.29**	**3.18**
27	5%	4.21	3.35	2.96	2.73	2.57	2.46	2.37	2.31	2.25
	1%	**7.68**	**5.49**	**4.60**	**4.11**	**3.78**	**3.56**	**3.39**	**3.26**	**3.15**
28	5%	4.20	3.34	2.95	2.71	2.56	2.45	2.36	2.29	2.24
	1%	**7.64**	**5.45**	**4.57**	**4.07**	**3.75**	**3.53**	**3.36**	**3.23**	**3.12**
29	5%	4.18	3.33	2.93	2.70	2.55	2.43	2.35	2.28	2.22
	1%	**7.60**	**5.42**	**4.54**	**4.04**	**3.73**	**3.50**	**3.33**	**3.20**	**3.09**
30	5%	4.17	3.32	2.92	2.69	2.53	2.42	2.33	2.27	2.21
	1%	**7.56**	**5.39**	**4.51**	**4.02**	**3.70**	**3.47**	**3.30**	**3.17**	**3.07**
40	5%	4.08	3.23	2.84	2.61	2.45	2.34	2.25	2.18	2.12
	1%	**7.31**	**5.18**	**4.31**	**3.83**	**3.51**	**3.29**	**3.12**	**2.99**	**2.89**
60	5%	4.00	3.15	2.76	2.53	2.37	2.25	2.17	2.10	2.04
	1%	**7.08**	**4.98**	**4.13**	**3.65**	**3.34**	**3.12**	**2.95**	**2.82**	**2.72**
120	5%	3.92	3.07	2.68	2.45	2.29	2.18	2.09	2.02	1.96
	1%	**6.85**	**4.79**	**3.95**	**3.48**	**3.17**	**2.96**	**2.79**	**2.66**	**2.56**

Table G*   Values of chi square (χ^2) at the 5% and 1% levels of significance.

Degrees of Freedom (*df*)	5%	1%
1	3.84	6.64
2	5.99	9.21
3	7.82	11.34
4	9.49	13.28
5	11.07	15.09
6	12.59	16.81
7	14.07	18.48
8	15.51	20.09
9	16.92	21.67
10	18.31	23.21
11	19.68	24.72
12	21.03	26.22
13	22.36	27.69
14	23.68	29.14
15	25.00	30.58
16	26.30	32.00
17	27.59	33.41
18	28.87	34.80
19	30.14	36.19
20	31.41	37.57
21	32.67	38.93
22	33.92	40.29
23	35.17	41.64
24	36.42	42.98
25	37.65	44.31
26	38.88	45.64
27	40.11	46.96
28	41.34	48.28
29	42.56	49.59
30	43.77	50.89

* Table G is abridged from Table IV of Fisher and Yates, *Statistical Tables for Biological, Agricultural, and Medical Research*, 6th edition, 1963, published by Longman Group Ltd., London (previously published by Oliver and Boyd, Edinburgh), by permission of the authors and publishers.

Table H* Values of T at the 5% and 1% levels of significance in the Wilcoxon signed-ranks test.

N_{S-R}	5%	1%
6	0	—
7	2	—
8	4	0
9	6	2
10	8	3
11	11	5
12	14	7
13	17	10
14	21	13
15	25	16
16	30	20
17	35	23
18	40	28
19	46	32
20	52	38
21	59	43
22	66	49
23	73	55
24	81	61
25	89	68

* Table H is adapted from Table I of F. Wilcoxon, *Some Rapid Approximate Statistical Procedures*, New York, American Cyanamid Co., 1949, with the permission of the publisher.

Table I* Values of q_α (the Studentized range statistic) at the 5% and 1% levels of significance.

df_{wg}	α	2	3	4	5	6	7	8	9	10
		\multicolumn{9}{c}{$k = Number\ of\ Means$}								
1	.05	17.97	26.98	32.82	37.08	40.41	43.12	45.40	47.36	49.07
	.01	90.03	135.00	164.30	185.60	202.20	215.80	227.20	237.00	245.60
2	.05	6.08	8.33	9.80	10.88	11.74	12.44	13.03	13.54	13.99
	.01	14.04	19.02	22.29	24.72	26.63	28.20	29.53	30.68	31.69
3	.05	4.50	5.91	6.82	7.50	8.04	8.48	8.85	9.18	9.46
	.01	8.26	10.62	12.17	13.33	14.24	15.00	15.64	16.20	16.69
4	.05	3.93	5.04	5.76	6.29	6.71	7.05	7.35	7.60	7.83
	.01	6.51	8.12	9.17	9.96	10.58	11.10	11.55	11.93	12.27
5	.05	3.64	4.60	5.22	5.67	6.03	6.33	6.58	6.80	6.99
	.01	5.70	6.98	7.80	8.42	8.91	9.32	9.67	9.97	10.24
6	.05	3.46	4.34	4.90	5.30	5.63	5.90	6.12	6.32	6.49
	.01	5.24	6.33	7.03	7.56	7.97	8.32	8.61	8.87	9.10
7	.05	3.34	4.16	4.68	5.06	5.36	5.61	5.82	6.00	6.16
	.01	4.95	5.92	6.54	7.01	7.37	7.68	7.94	8.17	8.37
8	.05	3.26	4.04	4.53	4.89	5.17	5.40	5.60	5.77	5.92
	.01	4.75	5.64	6.20	6.62	6.96	7.24	7.47	7.68	7.86
9	.05	3.20	3.95	4.41	4.76	5.02	5.24	5.43	5.59	5.74
	.01	4.60	5.43	5.96	6.35	6.66	6.91	7.13	7.33	7.49
10	.05	3.15	3.88	4.33	4.65	4.91	5.12	5.30	5.46	5.60
	.01	4.48	5.27	5.77	6.14	6.43	6.67	6.87	7.05	7.21
11	.05	3.11	3.82	4.26	4.57	4.82	5.03	5.20	5.35	5.49
	.01	4.39	5.15	5.62	5.97	6.25	6.48	6.67	6.84	6.99
12	.05	3.08	3.77	4.20	4.51	4.75	4.95	5.12	5.27	5.39
	.01	4.32	5.05	5.50	5.84	6.10	6.32	6.51	6.67	6.81
13	.05	3.06	3.73	4.15	4.45	4.69	4.88	5.05	5.19	5.32
	.01	4.26	4.96	5.40	5.73	5.98	6.19	6.37	6.53	6.67
14	.05	3.03	3.70	4.11	4.41	4.64	4.83	4.99	5.13	5.25
	.01	4.21	4.89	5.32	5.63	5.88	6.08	6.26	6.41	6.54
15	.05	3.01	3.67	4.08	4.37	4.59	4.78	4.94	5.08	5.20
	.01	4.17	4.84	5.25	5.56	5.80	5.99	6.16	6.31	6.44
16	.05	3.00	3.65	4.05	4.33	4.56	4.74	4.90	5.03	5.15
	.01	4.13	4.79	5.19	5.49	5.72	5.92	6.08	6.22	6.35

* Table I is adapted from Table 29 of Pearson, E. S., and Hartley, H. O. (eds.), *Biometrika Tables for Statisticians*. Vol. I (3rd ed.) Cambridge: Published for the Biometrika Trustees at the University Press, 1966, with the permission of the Biometrika Trustees.

Values of q_α (the Studentized range statistic) at the 5% and 1% levels of significance, *continued*.

					k = Number of Means					
df_{wg}	α	2	3	4	5	6	7	8	9	10
17	.05	2.98	3.63	4.02	4.30	4.52	4.70	4.86	4.99	5.11
	.01	4.10	4.74	5.14	5.43	5.66	5.85	6.01	6.15	6.27
18	.05	2.97	3.61	4.00	4.28	4.49	4.67	4.82	4.96	5.07
	.01	4.07	4.70	5.09	5.38	5.60	5.79	5.94	6.08	6.20
19	.05	2.96	3.59	3.98	4.25	4.47	4.65	4.79	4.92	5.04
	.01	4.05	4.67	5.05	5.33	5.55	5.73	5.89	6.02	6.14
20	.05	2.95	3.58	3.96	4.23	4.45	4.62	4.77	4.90	5.01
	.01	4.02	4.64	5.02	5.29	5.51	5.69	5.84	5.97	6.09
24	.05	2.92	3.53	3.90	4.17	4.37	4.54	4.68	4.81	4.92
	.01	3.96	4.55	4.91	5.17	5.37	5.54	5.69	5.81	5.92
30	.05	2.89	3.49	3.85	4.10	4.30	4.46	4.60	4.72	4.82
	.01	3.89	4.45	4.80	5.05	5.24	5.40	5.54	5.65	5.76
40	.05	2.86	3.44	3.79	4.04	4.23	4.39	4.52	4.63	4.73
	.01	3.82	4.37	4.70	4.93	5.11	5.26	5.39	5.50	5.60
60	.05	2.83	3.40	3.74	3.98	4.16	4.31	4.44	4.55	4.65
	.01	3.76	4.28	4.59	4.82	4.99	5.13	5.25	5.36	5.45
120	.05	2.80	3.36	3.68	3.92	4.10	4.24	4.36	4.47	4.56
	.01	3.70	4.20	4.50	4.71	4.87	5.01	5.12	5.21	5.30
∞	.05	2.77	3.31	3.63	3.86	4.03	4.17	4.29	4.39	4.47
	.01	3.64	4.12	4.40	4.60	4.76	4.88	4.99	5.08	5.16

Index